INTELLECTUAL PROPERTY PROTECTION IN VLSI DESIGNS

Intellectual Property Protection in VLSI Designs

Theory and Practice

by

Gang Qu
University of Maryland, U.S.A.

and

Miodrag Potkonjak
University of California, Los Angeles, U.S.A.

SPRINGER SCIENCE+BUSINESS MEDIA, LLC

A C.I.P. Catalogue record for this book is available from the Library of Congress.

ISBN 978-1-4419-5327-8 ISBN 978-0-306-48717-0 (eBook)
DOI 10.1007/978-0-306-48717-0

Printed on acid-free paper

Originally published by Kluwer Academic Publishers in 2003
Softcover reprint of the hardcover 1st edition 2003

Contents

List of Figures

List of Tables

To my parents, my wife, and my son. –Gang Qu

Acknowledgments

Intellectual property protection of hardware and software artifacts is of crucial importance for a number of dominating business models. Maybe even more importantly, it is an elegant and challenging scientific and engineering challenge. This book provides in detailed treatment of our newly developed constraint-based protection paradigm for the protection of intellectual properties in VLSI CAD. The key idea is to superimpose additional constraints that correspond to an encrypted signature of the designer to design/software in such a way that quality of design is only nominally impacted, while strong proof of authorship is guaranteed. Its basis is the Ph.D. dissertation of the first author. In addition, it also presents a few of the most recent research results from both authors and their colleagues.

We are grateful to our co-authors who greatly contributed to research presented in this book including Andrew Caldwell, Hyun-Jin Choi, Andrew Kahng, Darko Kirovski, David Liu, Stefanus Mantik, and Jennifer Wong. In addition, we would also like to thank a number of other researchers, including Jason Cong, Inki Hong, Yean-Yow Huang, John Lach, William Magione-Smith, Igor Markov, Huijuan Wang, and Greg Wolf for numerous advises and even more numerous helpful discussions.

We would also like to acknowledge Virtual Socket Interface Alliance for allowing us to include its document, "Intellectual Property Protection White Paper: Schemes, Alternatives and Discussion Version 1.1", as the appendix. Special thanks to Stan Baker, Executive Director of VSI Alliance, and Ian Mackintosh, author of the above document, for making this happen.

Finally, we would like to thank Pushkin Pari and Jennifer Wong for careful reading of the manuscript and for providing us invaluable feedback. We would like to express appreciation to our publishing editor, Mark de Jongh, for his help throughout this project. Any errors that remain are, of course, our own.

Gang Qu
College Park, Maryland
gangqu@glue.umd.edu

Miodrag Potkonjak
Los Angeles, California
miodrag@cs.ucla.edu

September 2002

Chapter 1

DESIGN SECURITY: FROM THE POINT OF VIEW OF AN EMBEDDED SYSTEM DESIGNER

I first observed the "doubling of transistor density on a manufactured die every year" in 1965, just four years after the first planar integrated circuit was discovered. The press called this "Moore's Law" and the name has stuck. To be honest, I did not expect this law to still be true some 30 years later, but I am now confident that it will be true for another 20 years.

—Gordon E. Moore

1. Introduction

According to the International Technology Roadmap for Semiconductors [169], there are now 42 million transistors on a chip, and this number is projected by Moore's Law to reach 400 million by 2005. With this ever-increasing chip capacity, it is expected that we can implement more complex systems on a single chip, which require longer design and verification cycle. Meanwhile, the time-to-market window keeps on shrinking due to the global competition and corporate cost cutting to design new products, particularly embedded systems[1]. System designer's design productivity increases, but at a much slower pace. This creates a design productivity gap between what can be built and what can be designed. To close this gap, we need a significant shift in design methodology, and at the center of this shift is the principle of design reuse. In this new design method, previously designed large blocks will be integrated into an ASIC (Application Specific Integrated Circuit) architecture which also includes new design blocks, representing true innovation on the part of the design team. Among the existing technical and non-technical barriers for reuse-based design methodology to thrive, intellectual property (IP) protection is a unique and one of the most challenging areas awaiting research breakthroughs.

What makes IP protection a unique challenge is the new reuse-based design environment. IP reuse forces engineers to cooperate with others and share their data, expertise, and experience. Design details (including the RTL HDL source codes) are encouraged to be documented and made public for better and more convenient reuse. The advances in the Internet and the World Wide Web play an important role as we have seen many web-based design tools emerging in the past few years that enable geographically separated design teams to cooperate. But at the same time, this makes IP piracy and infringement easier than ever. It is estimated that the annual revenue loss in IP infringement in IC (Integrated Circuit) industry is in excess of $5 billion. As summarized in [105], the goals of IP protection include: enabling IP providers to protect their IPs against unauthorized use, protecting all types of design data used to produce and deliver IPs, detecting and tracing the use of IPs.

In this chapter, we briefly review the reuse-based design methodology and discuss the need of protection techniques in embedded system design and VLSI (Very Large Scale Integration) CAD (Computer Aided Design). We will present a couple of small examples to illustrate our newly developed constraint-based IP protection techniques. We conclude with an overview of the proposed IP protection paradigm that consists of watermarking, fingerprinting, and copy detection.

2. Intellectual Property in Reuse-Based Design

2.1 The Emergence of Embedded Systems

The notion of *embedded systems* is first used for certain military applications, for instance, weapon control or, in a broader sense, military command, control and communication systems. Later on, people call "electronic systems embedded within a given plant or external process with the aim of influencing this process in a way that certain overall functional and performance requirements are met", embedded systems [96]. We have seen embedded systems emerging in the past decade mainly due to the thriving Internet. Conventional stand-alone embedded systems are now increasingly becoming connected via networks. Embedded systems, as a combination of hardware and software that perform a specific function, now can be found almost everywhere:

- at home: appliances like toaster, microwave, dish washer, answering machine, washing machine, drier, ...
- in the office: equipments like printer, fax machine, scanner, copier, ...
- in our daily life: devices like cellular phone, personal digital assistants, cameras, camcorders, ...
- in automobiles, planes, and rockets: parts like fuel injection, anti-lock brakes, engine control, ...

Many of these devices are not new, however, they are normally isolated until the Internet makes them network-centered. As a result, it becomes possible to have wireless communications, multimedia applications, interactive games, TV set-top boxes, video conferences, video-on-demand, etc. In 1997, the average U.S. household had over 10 embedded computers, not to mention the automobile, which has more than 35 at the end of year 2000. Demand for embedded system designers is large, and is growing rapidly. For example, every year, there are more than 5 billions embedded systems sold in the world, comparing to less than 120 millions general purpose systems. According to the International Data Corporation, by the year 2002, the Internet appliance itself will see a larger market than PC market.

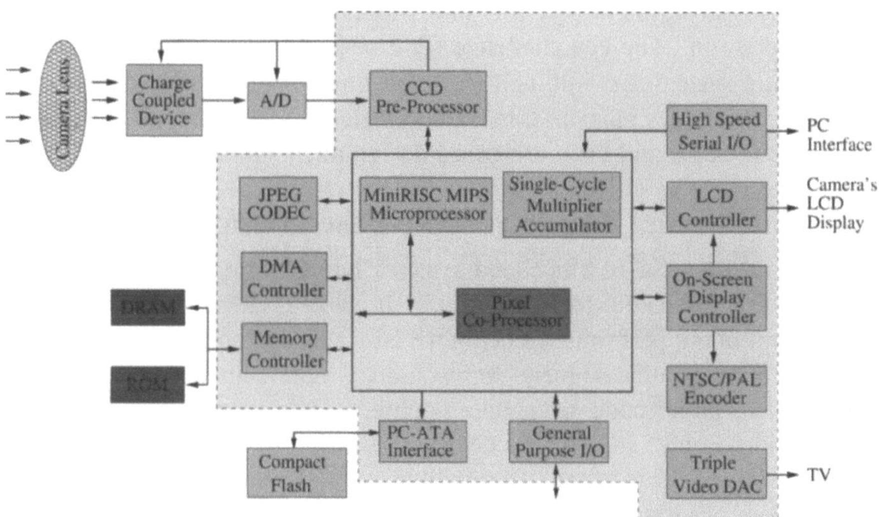

Figure 1.1. Block diagram of the DCAM-103 digital camera (redrawn from the website of LSI Logic Corp.).

Figure 1.1 shows the architecture of one such embedded system, the DCAM-103 digital camera from LSI Logic Corp. (*http://www.lsilogic.com/*). It is a highly integrated single-chip processor that processes still images: preview, capture, compress, store, and display. LSI Logic CW4003 processor core is engineered to provide efficient processing of digital images. A pixel co-processor enables fast processing of edge enhancement, image resizing, color conversion, pixel interpolation, etc. The multiplier accumulator assists certain digital signal processing. The CCD (Charge Coupled Device) pre-processor reads the digital representation created by the CCD and processes it to produce color images. The JPEG codec compresses/decompresses images. DMA and memory controllers control the access to local image memory. Other devices ensure the integration with peripherals, printers, computers, TVs, scanners, and so on.

The system implements single functionality (i.e., digital still image processing: captures, compresses and stores frames; decompresses and displays frames; uploads frames.). Its design is tightly constrained featuring low cost, small size, high performance, and low power consumption.

Unlike the general purpose systems (workstations, desktops, and notebook computers), which are designed to maximize the number of devices sold and thus are designed to meet a variety of applications, embedded systems have their own common characteristics. As we have seen in the case of the digital camera, first, they are usually single-functioned; secondly, there exist tight design constraints; and thirdly such systems deal with reactive and real-time applications. The design constraints include size, performance, power, unit cost, non-recurring cost, flexibility, time-to-market, time-to-prototype, correctness, safety, and so on. The key challenge for embedded system design is how to implement a system that fulfills the desired functionality and simultaneously optimizes various design metrics in a timely fashion. One of the most successful answers is IP reuse and the reuse-based design methodology.

2.2 Intellectual Property Reuse-Based Design

The rapid increase of embedded systems has brought an historic technological change in the electronics industry. It challenges the system designers' assumptions about performance being the No. 1 design bottleneck. Other factors are climbing into designers' top wish list: more complex processors and architectures, larger code size, more complicated functionalities, less power consumption, lighter and smaller devices, shorter time-to-market, lower cost, etc. Meanwhile, silicon capacity is doubling every 18 months thanks to the rapid advancement of fabrication technologies. Now it is possible to build systems on a single chip of silicon (System-On-a-Chip) under $0.18\mu m$ with a couple of millions of gates. This provides the necessary condition for building complex but small-size systems for the new applications. However, design team's expertise and productivity as well as their design tools cannot grow at the same pace. As the design complexity goes up, we should expect longer design cycle. But what we get in reality is the time-to-market pressure. The gap between silicon capacity and design productivity seems to be widening at an even greater pace, slowing the growth of the semiconductor industry.

As a result, companies will be forced to specialize and focus on the things that they do best, and partner with others for the necessary components to bring the whole system to market in a competitive time frame. This leads to the concepts of design reuse and IP based design methodology. In the past few years, organizations such as VSIA (Virtual Socket Interface Alliance) and VCX (Virtual Component Exchange) have attracted large number of companies in order to make SOC design a practical reality by mixing and matching the IPs. For example, more than 200 leading systems, semiconductor, IPs and EDA (Electronic Design Automation) vendors have joined VSIA which is working on IP implementation, interface, protection, testing, and verification among other challenges for IP reuse. VCX has launched a number of development working groups to define trading standards for IP exchange.

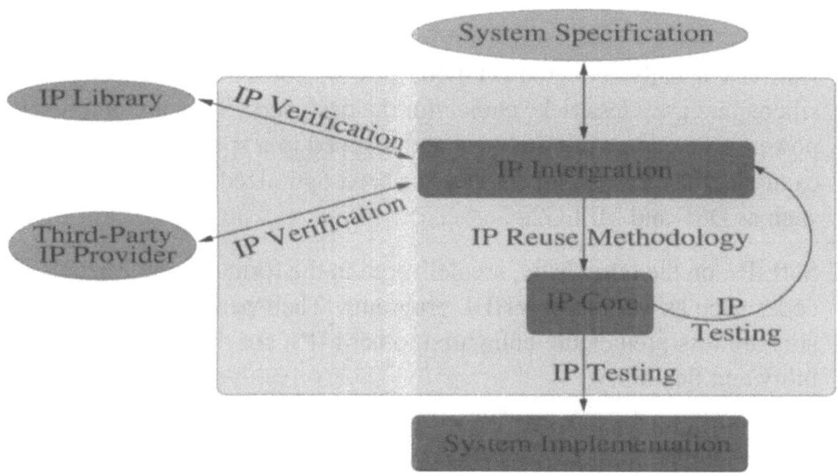

Figure 1.2. Intellectual property reuse-based design flow.

Figure 1.2 depicts the global design flow based on IP reuse. With the system specification, the designers will take the necessary virtual components (IPs) from the IP library and the third-party IP providers. The IP library can be internal or external. An IP verification process is required for external IPs and IPs from third-party IP providers. Then designers can exploit the reuse methodology to build the core in a much more efficient way than design-from-scratch. After IP testing is accomplished, this design can be added to the internal IP library for later use and will have market value.

We can see this for the design of DCAM-103 digital camera (Figure 1.1) where the design objective is to process typical digital still images. According to the corresponding requirements, technologies in the previous DCAM series (e.g, the LSI Logic CW4003 processor core and the pixel co-processor), JPEG codec, and other additional logic have been selected from the IP library to integrate the core. Once the core has been tested, it is included in the (internal) IP library for future reuse, and the DCAM development system (the DCAM-103 device, demonstration hardware, DCAM reference software, and the optional FlashPoint Technology's Digita operating environment) is built around the core to provide customers the flexibility of integrating with their own IP to ensure different solutions.

Intellectual property typically refers to products of the human intellect, such as ideas, inventions, expressions, unique names, business methods and formulas, mask works, information, data, and know-how. In the EDA society, it refers to pre-designed blocks, also known as IP blocks, cores, system-level blocks, macros, megacells, system level macros, or virtual components. The most valuable asset of such IPs are the ideas, concepts, or algorithms that make such hardware and/or software functional.

IPs can be put into many different categories. VLSI design IPs are either *hard* or *soft*.

- Hard IPs, usually delivered as GDSII files, are cores that have been proven in silicon and are a less risky choice for the designers. They are optimized for power, size, and/or performance and mapped to a specific technology. For example, the physical layout that has been optimized for a specific process such as DSP and MPEG2.

- Soft IPs, on the other hand, are delivered in the form of synthesizable HDL codes such as Verilog or VHDL programs. Their performance, power, and area are less predictable compared to hard IPs, but they offer better portability and flexibility.

A compromise between hard IP and soft IP is the so-called *firm IPs* such as placement of RTL blocks, fully placed netlist, or guidance for physical placement and floorplanning. Firm IPs normally, although not mandatory, include synthesizable RTL HDL files. In [5], *physical libraries* are defined to be the physical building blocks that include such things as memory, standard cells, and datapaths; *board libraries* are the IPs such as LSI, MSI, and gates; *software libraries* are fixed function in embedded software targeted to a specific microprocessor such as a RTOS or FTP.

There are many interpretations on the value of IPs. For example, in [156], IP's value is considered as the measure of the utility or profitability that ownership of IP brings to the enterprise. IP's value is measured both quantitatively and qualitatively. Quantitative measurements reveal how much profit and in what direction (increase vs. decrease) IP provides value. Qualitative measurements provide a sense of how the value is provided. Further discussion on the value and management of IPs can be found in a white paper issued by VSIA's IP Protection Development and Working Group, which is available at *http://www.vsi.org*.

IPs provide designers with reusable building blocks that can be used in future products. As a result, designers can spend more time focusing on the proprietary portions of a design rather than starting from scratch. This IP reuse-based design methodology has been proven to be the most powerful design technology innovation to increase design productivity. Figure 1.3 depicts the major design technology innovations and their impact to design productivity since RTL design methodology originated in 1990[169]. Clearly design reuse has

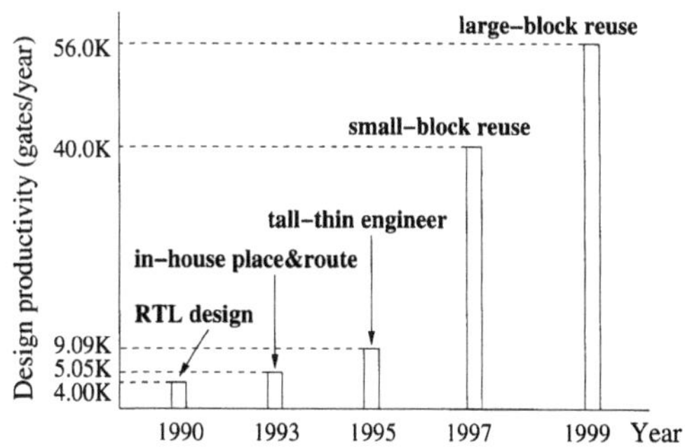

Figure 1.3. Design technology innovations and their impact to design productivity.

made the greatest contribution in improving the design productivity. There are also a number of successful stories of design reuse: Hitachi has reduced the number of late projects from 72% to 7% in four years; HP has shortened its products' time-to-market by a factor of 4 while reduced error rate by a factor of 10; Toshiba has improved its productivity 3 times in nine years.

The intellectual property reuse in the reuse-based design methodology is different from the reuse of devices such as decoders, multiplexers, registers, and counters to produce large systems. First, the level of integration is different. Reusable IP blocks consist of tens of thousands to millions of gates. Second, the complexity of reuse is different. IP functional verification becomes much more complicated, let alone the problems of making necessary modifications, handling analog/mixed signals and on-chip buses, conducting manufacturing related test and so on. Third, design target is different. In reuse-based design, design for reuse becomes a critical design objective for all designs.

As suggested in the "Reuse Methodology Manual for System-On-A-Chip Designs"[84], the process of integrating IPs and doing physical chip design can be broken into the following steps:

- Selecting IP blocks and preparing them for integration.
- Integrating all the IP blocks into the top-level RTL.
- Planning the physical design.
- Synthesis and initial timing analysis.
- Initial physical design and timing analysis, with iteration until timing closure.
- Final physical design, timing verification, and power analysis.
- Physical verification of the design.

There are many technical/non-technical issues need to be addressed for IP market to flourish: friendly interface between IP provider and IP user, design-for-test, design-for-reuse, easy-to-use, easy-to-verify, IP standardization, and rules for IP exchange. IP reuse is based on information sharing and integration. Therefore pirates will also have much easier access to the IPs, and IP protection becomes one of the key enabling techniques for industrial strength reuse-based synthesis.

2.3 Intellectual Property Misuse and Infringement

New technologies bring new applications and business models, however, they also find themselves the target for misappropriation almost immediately. Consider only the software industry, according to a recent survey commissioned by the Business Software Alliance (*http://www/nopiracy.com*) and the Software and Information Industry Association (*http://www.siia.net*), more than 38% of all software used in the world is illegally copied. This causes a $11 billion revenue loss in 1998, more than $12 billion in 1999, and a total of more than $59 billion during the past five years, leaving alone the consequences of fewer jobs, less innovation, and higher costs for consumers.

Further difficulty grows in hardware misuse and infringement. The growing black market business of manufacturing pirated hardware is flooding markets with cheap and surprisingly reliable alternatives to the expensive big brand names like Intel. As the time-to-market pressure drives intellectual property into the center of several trends sweeping through today's electronic design automation (EDA) and application specific integrated circuits (ASIC) industries, IP becomes a very lucrative target for pirates. Meanwhile, the growth and full utilization of the Internet, combined with revolutionary developments in the World Wide Web, have made (Internet) piracy much easier than ever. Various methods have been used by IP pirates to offer and distribute pirated IPs: E-mail, FTP, news groups, bulleting boards, Internet relay chat, direct/remote site links, and much more.

We name a few law suits involving IP infringement from a fast growing list: Sega Enterprises Ltd. v. Accolade Inc. in 1992 for the game cartridges [2], Intel Corp. v. Terabyte Intern. Inc. in 1993 for Intel trademark infringement[3], Apple Computer Inc. v. Microsoft Corp. in 1994 for the use of Apple's GUI[4], Cadence Inc. v. Avant! Corp. in 1995 for the copy of source code[5], Sony Inc. v. Connectix Corp. in 1999 for the copy of Sony's copyrighted BIOS[6], and the lawsuit against Napster, Inc. by a number of major recording companies in 2001[7].

Besides the numerous federal and state laws and regulations on intellectual property (copyright, trademark, patent, trade secret, antitrust, unfair competition, and so on) infringement, there are technical efforts (often referred as *self protection*) directly from the IP creators to keep their IPs beyond the reach of pirates. Watermarking or data hiding is one of the most widely used techniques. In essence, watermarking intentionally embeds digital information into the IP for purposes such as identification and copyright. Such information could be the author's name, company name or other messages highly related to the owner and/or the legal users of the IP. If necessary, this information can be used in court to prove the authorship of the IP or the legal users entitled to distribute copies.

For one type of IP (e.g. text, image, audio, video), watermark can be easily put into the digital content as minute changes. Although this alters the original IP, it remains useful as long as the end users cannot tell the difference. For example, in the context of plain text watermarking, various techniques have been developed to utilize inter-sentence space, end-of-line space, inter-word space, punctuation, synonyms, and many other features. Combined with modern cryptographic tools (e.g., encryption, public-key, private key, pretty good privacy), this method is proven very successful in providing protection for data and information.

The IP we discuss here is of a quite different type in the sense that the IP's utility relies on its correct functionality. The biggest challenge is how to hide signatures without changing the functionality. We have seen serial numbers being etched on the chip, redundant code being left in the source code, variable naming and programming styles also being used as evidence of the authorship, and so on. However, all these protection methods are vulnerable to attacks: serial numbers can be removed or changed, useless portion of the code can be detected and deleted, variables can be renamed, $\cdots$. The effectiveness of such protection is way lower than what we have been seeking.

One of the reasons that make these efforts not that successful is that the protection process is handled independently of the design and implementation of the IP. To add protection on top of an already functioned IP, the IP designers do not have much advantages over the attackers. On the contrary, they usually do not possess the expertise that professional attackers have and are not well aware of how powerful the attacking tools can be. For instance, the Intel 80386 has been successfully reverse engineered in a university lab in 1993. It took only six instances of the chip and less than two weeks[8].

As a conclusion, it is too late to have protection as the last phase of IP design. Instead, protection has to be done simultaneously with the design and implementation process, when the designer has all the controls that nobody later on can gain from a finalized IP. The constraint-based IP protection is based on this observation.

3. Constraint-Based IP Protection: Examples

We illustrate the constraint-based intellectual property protection techniques by several examples: the Boolean satisfiability (SAT) problem, FPGA design for the digital encryption standard (DES) benchmark, the graph vertex coloring (GC) problem, and design of the 4th order continued fraction infinite impulse response (CF IIR) filter.

3.1 Solutions to SAT

In the Boolean satisfiability problem (*SAT*), we have a formula of boolean variables and want to decide whether there is a truth assignment (true or false) for each of the variables such that the formula is true. For example, $\mathcal{F} = x_1' \cdot (x_1 + x_2) \cdot (x_1' + x_2)$ is satisfiable by assigning $x_1 = 0$ (for false) and $x_2 = 1$ (for true). However, formula $x_1 \cdot (x_1' + x_2) \cdot (x_1' + x_2')$ cannot be satisfied no matter which values we assign to variables x_1 and x_2. SAT is well-known as the first problem shown to be NP-complete, and the starting point for building the theories of NP-completeness[61]. Because of its discrete nature, SAT appears in many contexts in the field of VLSI CAD, such as automatic pattern generation, logic verification, timing analysis, delay fault testing and channel routing. Many heuristics have been developed to solve SAT problem due to its complexity and importance[173, 107]. Solution(s) to a hard SAT problem is definitely a piece of IP that can be easily misused. For instance, once the satisfying assignment is announced, everyone who makes use of it can claim he/she finds the solution by himself/herself. The real IP owners cannot distinguish themselves and fail to protect this piece of IP.

Our "simple" mission is to solve the SAT instance in such a way that we are able to demonstrate that we solve it. The technique we use here modifies the original SAT formula to force the solution we get have certain structure. This structure contains information (signature or watermark) corresponding to our authorship. We take advantage of one interesting feature of SAT: there may exist more than one truth assignments if the formula is satisfiable. Consider the following formula $\mathcal{F}$ of 13 variables:

$$(x_3 + x_4)(x_1' + x_3')(x_9' + x_8')(x_3 + x_{11})(x_{13}' + x_{12}')(x_1 + x_2 + x_3')(x_6 + x_8 + x_9)$$
$$(x_1 + x_3 + x_7)(x_4 + x_{11} + x_{10})(x_1' + x_2' + x_3')(x_{12} + x_5' + x_2)(x_7' + x_6 + x_4 + x_5)$$
$$x_2 \cdot x_5 \cdot (x_2' + x_3 + x_5 + x_9)(x_3 + x_5 + x_7' + x_{12})(x_6 + x_4' + x_8 + x_{11} + x_{13}')$$

an exhaustive search indicates that $\mathcal{F}$ is satisfiable and there are *256* distinct satisfying assignments. Now we encode a plain English message into new clauses using a simple case-insensitive scheme: letters "a - z" are mapped to $x_1, x_1', \cdots, x_{13}, x_{13}'$ alphabetically. For example, word "*red*" is encoded as $x_9' + x_3 + x_2'$ and the phrase "*A red dog*" is translated to $x_1(x_9' + x_3 + x_2')(x_2' + x_8 + x_4)$. After embedding the message "*A red dog is chasing the cat*", we add seven extra clauses, $x_1(x_9' + x_3 + x_2')(x_2' + x_8 + x_4)(x_5 + x_{10})(x_2 + x_4' + x_1 + x_{10} + x_5 + x_7' + x_4)(x_{10}' + x_4' + x_3)(x_2 + x_1 + x_{10}')$ to $\mathcal{F}$. Only *12* of the previous 256 truth assignments can satisfy these seven extra constraints. The solution we find will be guaranteed to be one of these 12.

Figure 1.4 is a Java demonstration showing this watermarking technique together with others which can be selected from the panel on the upper left corner. In the next panel, user can input signature in plain text. In the middle is the watermark key that converts the signature into SAT clauses shown at the

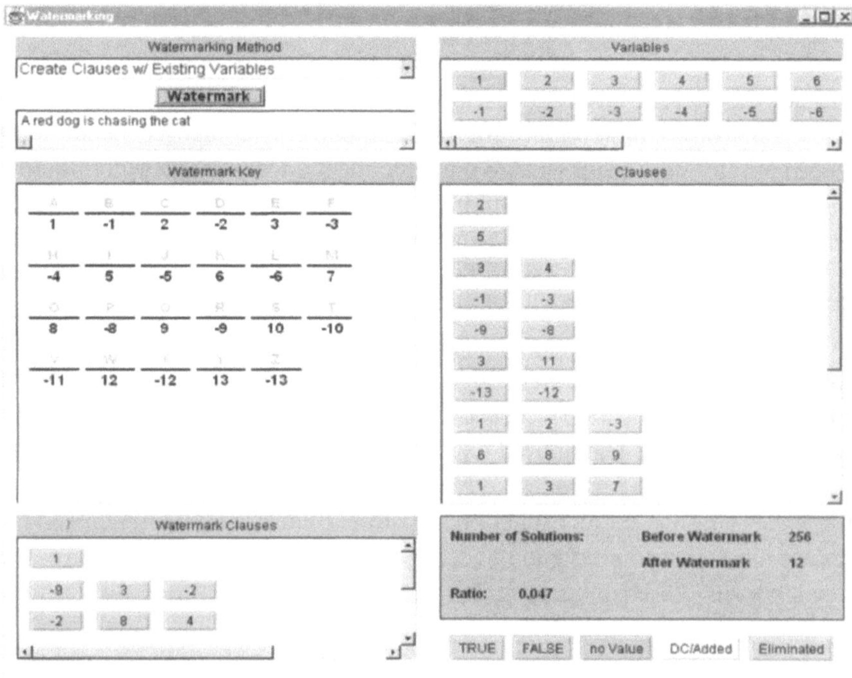

Figure 1.4. A Java GUI for watermarking the Boolean Satisfiability problem.

lower left panel. The right part describes the SAT instance and its solution. The "Variables" panel indicates the value of each variable in a given solution. Each row of the "Clauses" panel corresponds to a clause with the satisfied literal marked in pink and unsatisfied literal in green. As we can see, for a satisfiable instance, each row has at least one literal marked in pink. The blue (shaded) area gives the numbers of solutions before and after watermarking, as well as their ratio. This ratio quantitatively measures the uniqueness or the strength of the watermark. Smaller ratio implies stronger watermark.

Let us call this augmented SAT formula $\mathcal{F}'$, we observe that any solution to $\mathcal{F}'$ will have the following two properties: (i) it also makes the original formula $\mathcal{F}$ true; and (ii) it satisfies the above seven additional clauses. For any of these solutions, we claim that the likelihood of someone else finds this particular solution is $\frac{1}{256} \approx 0.39\%$, comparing to the chance of $\frac{1}{12} > 8.3\%$ for us. The odd is about 1:21, which is the strength of the watermark. For large SAT instances with hundreds of variables, this odd can be as small as 1:1,000,000 and provides a convincing proof for the authorship. More issues on protecting SAT solutions will be discussed in later chapters and can be found in [27, 133, 135].

3.2 FPGA Design of DES Benchmark

A field programmable gated array (FPGA) is a VLSI module that can be programmed to implement a digital system consisting of tens or hundreds of thousands of gates. It allows the realization of multi-level networks and complex systems on a single chip. An FPGA module is composed of an array of configurable logic blocks (CLB), interconnection points, and input/output blocks. This fixed standard structure of FPGA provides flexibility but leaves some CLBs and switches unused when being customized for a particular system. The non-trivial FPGA design task is how to implement a desired circuit using the minimal area of FPGA.

As demonstrated in [98], the FPGA design can be protected by embedding a secure and transparent watermark. In the proposed method (being applied to the Xilinx XC4000 architecture), each CLB contains two flip-flops and two 16x1 lookup tables (LUT). The unused CLBs are utilized to hide signatures. More specific, each free LUT encodes 16 bits of information; the netlist is modified, while preserving the correct functionality, to put constraints to the CLBs; the latter are then incorporated into the design with unused interconnection points and neighboring CLB inputs to further hide signatures.

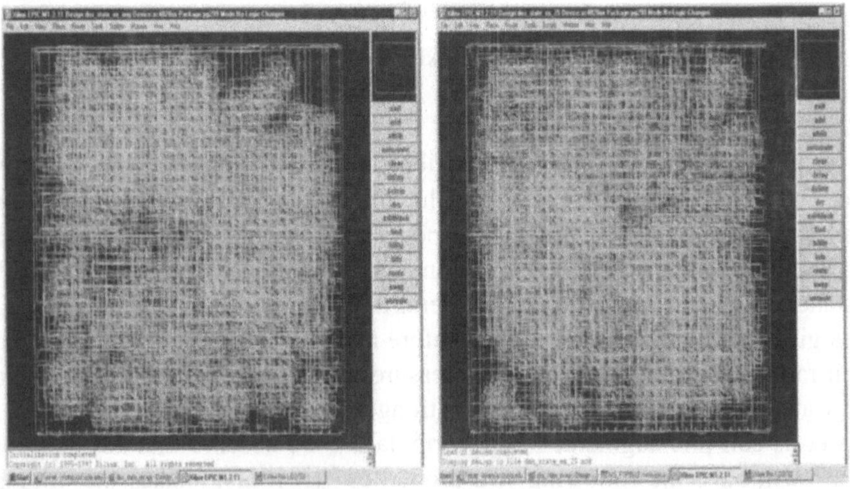

Figure 1.5. Layout of the DES benchmark without watermark(left) and the one with a 4768-bit message embedded (right).

This approach has been evaluated on the digital encryption standard (DES) design, a MIPS R2000 processor core, and a reconfigurable automatic target recognition system. In all the original physical layout of these systems, not only the entire LUTs and interconnections are not used, the place and route tools are not able to pack logic with optimal density as well. Therefore, it is possible to embed watermark by utilizing these free spaces without introducing area overhead. Figure 1.5 is the example of DES layouts. On the left is the original layout of the design. On the right is the design with an embedded signature of 4,768 bits. Notice that the original placement does not achieve optimal logic density. Instead, unused CLBs are dispersed throughout the design. Interestingly, timing analysis shows that there is actually no timing degradation in this case. In most of other experiments, the timing degradation is small or even negative, which means performance improvement.

3.3 Graph Coloring and the CF IIR Filter Design

As the final example, we show the NP-hard graph vertex coloring (GC) problem and one of its numerous applications in system design. This problem asks for a coloring of the vertices in a undirected graph with as few colors as possible, such that no two adjacent vertices (i.e., nodes that are connected by an edge) receive the same color. To protect the solution, we build a more constrained graph (by introducing additional edges) and color it instead of the original graph. The selection of such edges defines the encoding scheme. Similar to the SAT problem, we use a simple message encoding scheme to illustrate the watermarking technique, in which each letter of a given 4-letter message is encoded as an edge between a pair of unconnected vertices.

Considering a 19-node graph shown in Figure 1.6, we identify all the unconnected pairs (e.g., $(1,2),(1,3),(2,15),\ldots$) and sort them by the ascending order of the first and then the second vertices. Then each letter "A-Z" and "-" is encoded as one of these pairs alphabetically. The table on the right side of Figure 1.6 shows this encoding scheme. An entry with a solid (red) dot means the two vertices, whose indices coincide with this entry, are connected in the original graph. For example, the dot in the first row and sixteenth column says nodes 1 and 16 are connected. Based on this table, the message *UCLA* is translated to four edges: $(2,9),(3,4),(6,12),$ and $(8,13)$. These edges are added to the graph before we color it. The middle section of Figure 1.6 shows this and an obtained solution. We can see that it is quite different from the solution we have on top, which is obtained by a greedy searching strategy starting from a clique of size five (vertices 1, 12, 14, 16, and 18). The bottom figure is the result with message *VLSI* embedded.

Now we show how this technique can be applied for the protection of embedded system design. Figure 1.7 is the design of the 4th order continued fraction infinite impulse response filter, a very popular one used in embedded systems. As shown in the datapath (top of Figure 1.7), we implement it using one multiplier, one adder, and five registers. The ten control steps are repeated in an infinite loop. The table on the top left of Figure 1.7 shows that at each control step, how the nineteen variables are stored in the registers. One major concern of this design is to minimize the number of registers, which is the reg-

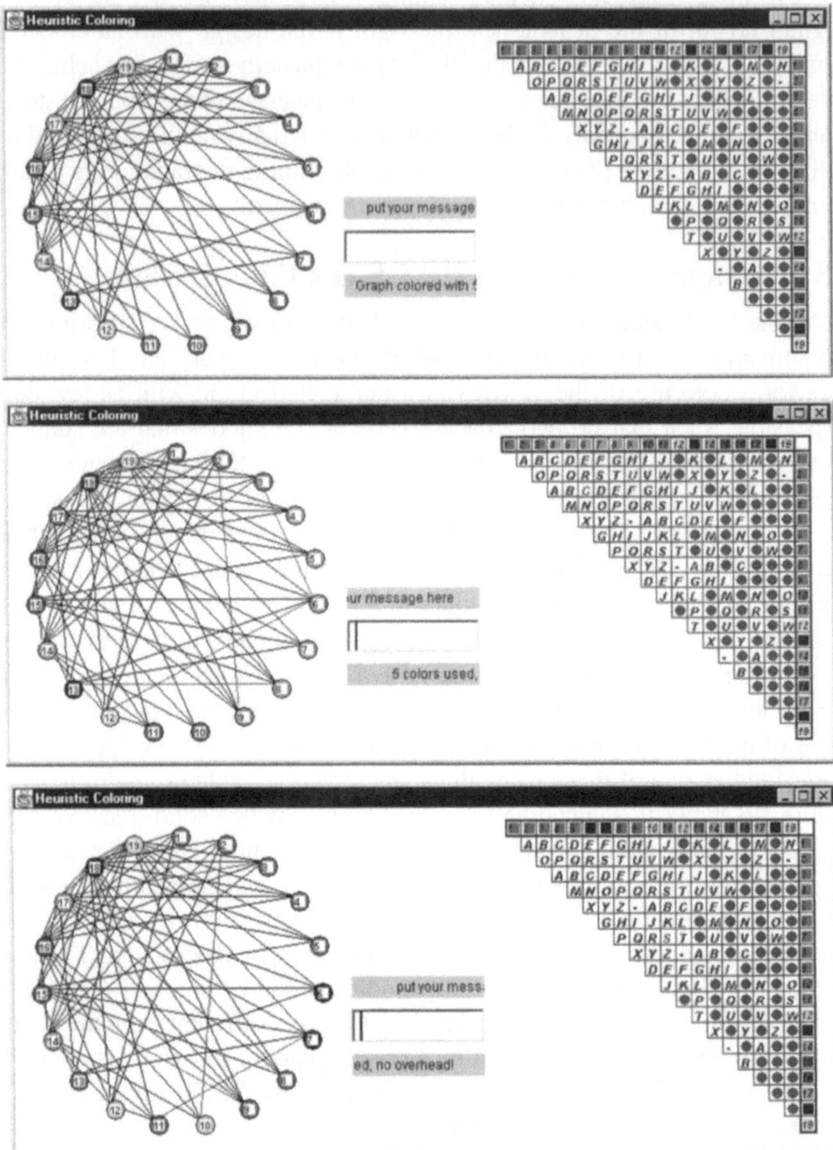

Figure 1.6. GUI for watermarking solutions to the graph coloring problem. (top: the greedy 5-color solution to the original graph; middle: a 5-color solution with message *UCLA* embedded; bottom: a 5-color solution with message *VLSI* embedded.).

ister allocation problem that is equivalent to the GC problem. From the control data flow graph (CDFG) and the scheduled CDFG, we observe that at sev-

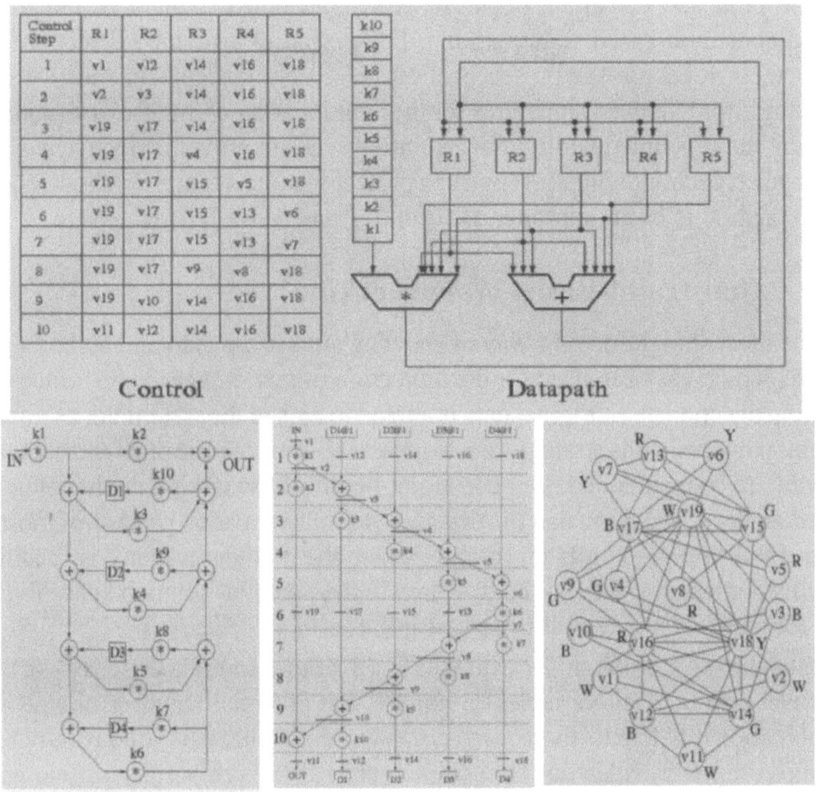

Figure 1.7. Design of the 4th order CF IIR filter with watermark. (top: control and datapath of the design implementation; bottom left: control data flow graph; bottom middle: scheduled CDFG; bottom right: colored interval graph.).

eral control steps, we need the values of five variables (For example, variables $v_1, v_{12}, v_{14}, v_{16}$, and v_{18} at step 1). This leads to the conclusion that at least five registers are required to enable high performance. In the corresponding interval graph (bottom right of Figure 1.7), this results in a clique of size five. In this implementation, we have embedded "A7" in ASCII which is extremely difficult to detect without knowing the rules for encoding. As one piece of evidence, we see that variable v_1 is assigned to register R1, while v_3 to R2. However, this is not necessary from the original constraints (from the scheduled CDFG, we see that v_1 and v_3 never alive in the same control step, which means that they may be assigned to the same register). It happens in our solution because an extra edge between v_1 and v_3 has been added in the interval graph to encode a bit 1, the most significant bit in 10000010110111 which is the ASCII code for "A7".

4. Constraint-Based IP Protection: Overview

The proposed constraint-based IP protection consists of three integrated parts: constraint-based watermarking, fingerprinting, and copy detection. Its correctness relies on the presence of all these components. In short, watermarking aims to embed signatures for the identification of the IP owner without altering the IP's functionality; fingerprinting seeks to provide effective ways to distinguish each individual IP users to protect legal customers; copy detection is the method to catch improper use of the IP and prove IP's ownership.

4.1 Constraint-Based Watermarking

The most straightforward way of showing authorship is to add author's signature, which has been used for the protection of text, image, audio, video and multimedia contents. Original data is altered to embed the watermark as minute errors. Obviously this strategy fails to protect IPs that require their correct functionality to be maintained. Our constraint-based watermarking methodology is based on the observation that the design and implementation process of most of such IPs is similar to problem solving, where the problem instance is specified as constraints and we are asked to search in the potential solution space to find one (or more) that meets all these constraints.

Take the SAT problem for example For a simple formula $\mathcal{F} = x_1' \cdot (x_1 + x_2)$ over two boolean variables x_1 and x_2, the potential solutions are all the combinations of 0/1 to these two variables; each clause is a constraint (for example, $x_1 + x_2$ rules out the assignment of 0 to both variables); we want to find a truth assignment to meet all the constraints (i.e., make all the clauses true), or show that such assignment does not exist in which case the formula is unsatisfiable. Any attempt of modifying the constraints may result in an incorrect solution: changing $x_1 + x_2$ to $x_1 + x_2'$ will guide the SAT solver to report the solution $x_1 = 0, x_2 = 0$, which does not satisfy the original formula $\mathcal{F} = x_1' \cdot (x_1 + x_2)$.

Constraint-based watermarking technique encodes signature as additional constraints, adds them into the problem specification and solve this more constrained problem instead of the original problem. Figure 1.8 illustrates this idea in system design process. In the traditional design process Figure 1.8(a), a designer simply uses the synthesis tools to obtain the best possible final design that meets all and only the initial specification. Since the final design satisfies nothing else but the given initial design constraints, the designer has no way to prove his authorship of this piece of IP. Being aware of the potential piracy, a more careful designer will embed his signature into the final design so that he can claim his authorship once the piracy occurs (Figure 1.8(b)). With the given initial design specification, the designer builds a watermarking engine which takes the design specification and designer's signature as input and returns the

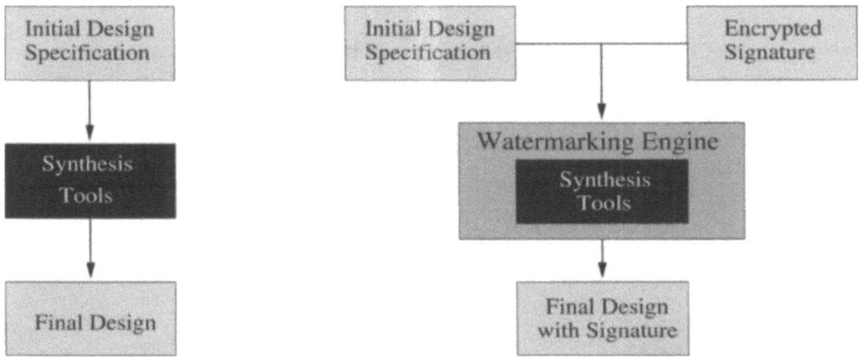

(a) Traditional design process. (b) New design process with watermarking.

Figure 1.8. Constraint-based watermarking in system design process. (left: traditional design flow; right: new design flow with watermarking process.).

final design. Inside the watermarking engine, the signature is translated into additional design constraints that the final design will satisfy as well. Notice that the satisfaction of these extra constraints is not necessary for a valid final design, so the designer can prove his authorship by showing the unlikelihood that this happens.

4.2 Fingerprinting

The goal of fingerprinting is to protect innocent IP users whenever IP misuse or piracy occurs. It is clear that to enable this, assigning different users distinct copies of the IP becomes necessary. One practical question is how to generate large amount of solutions efficiently. Figure 1.9 shows two of the protocols that we develop to answer this question: iterative fingerprinting technique and the constraint manipulation technique.

In iterative fingerprinting (Figure 1.9(a)), the original problem instance (usually large and expensive to solve) is solved once to obtain a seed solution; then a sub-problem of smaller size is generated based on the seed solution and the original problem; this small problem is solved again and we are able to get only a solution to the sub-problem, which is normally a partial solution to the original problem; this sub-solution is combined with the seed solution to build a new solution and will be served as the new seed solution in the next iteration. The cost for getting a new solution is much less than that for the original solution due to the fact that the problem's complexity decreases fast as we cut the size of the problem.

An even better solution in terms of run-time saving is the one based on constraint manipulation (Figure 1.9(b)). An augmented problem is derived

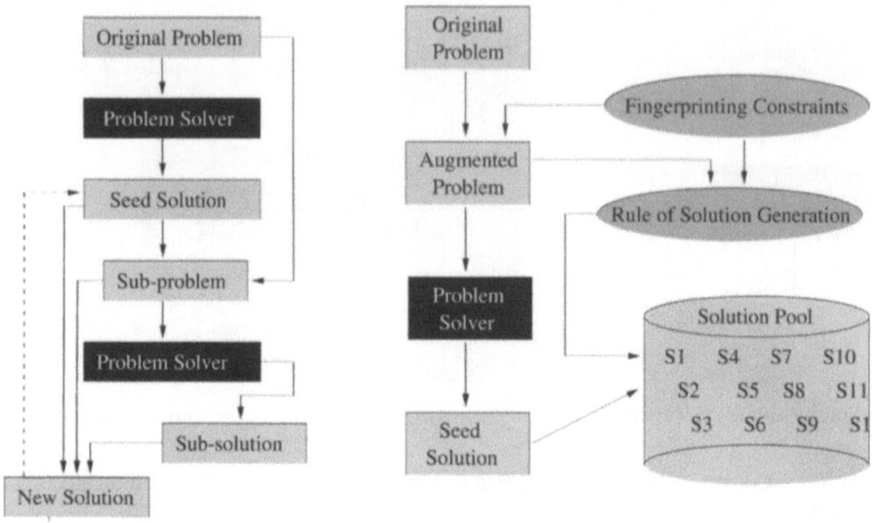

(a) Iterarive fingerprinting. (b) Fingerprinting by constraint manipulation.

Figure 1.9. Fingerprinting in system design process. (left: iterative fingerprinting technique; right: constraint addition based fingerprinting technique.).

from the original instance by adding constraints and then solved to get the seed solution. These constraints are selected such that the resulting seed solution will be well structured. According to the added fingerprinting constraints and the augmented problem, a set of rules is set up for creating new solutions from the seed solution. Since the solution generation process only involves this set of rules (which normally are all quite simple) and the seed solution, the problem solver will not be called again. The only run-time overhead comes from solving the more constrained augmented problem to get the seed solution.

As the basic idea of iterative fingerprinting technique comes from the iterative improvement approach for finding solutions to hard optimization problems, it is particular effective for optimization problems (e.g, partitioning, graph coloring, standard-cell placement.). The constraint addition method is generic, however, it is non-trivial to find such fingerprinting constraints and sometimes this may introduce non-negligible degradation in the solution's quality.

4.3 Copy Detection

Copy detection is an important part of our constraint-based IP protection paradigm. Without an effective copy detection method, all the previous efforts in watermarking and fingerprinting are in vain. Even when IP infringement and suspects are found, we cannot do much if we are unable to recover the watermark or fingerprint.

Complementary to watermarking and fingerprinting techniques, copy detection techniques aim to discover the hidden signature in a piece of IP. Suppose that the marks are embedded into the IP as additional constraints, we need to verify the existence of these constraints and show its connection with our signature[8]. However, most of these verification process are hard. Take the graph coloring problem we have discussed earlier for example. Since the watermarking technique depends on the ordering of the vertices[9], potentially every permutation of the vertices has to be checked which makes the run time goes up exponentially. Even worse is the case when the watermarked graph is embedded into a larger graph, then the task of finding the embedded marks becomes the well-known NP-complete graph isomorphism [61]. Unfortunately, this scenario happens in real life when a stolen IP is used to build another IP.

We argue that to assure fast detection, the watermark/fingerprint must be hidden behind certain parts of the problem with rather unique structure that are difficult to be altered. We call this methodology *watermarking for copy detection* or *detection-driven watermarking*. Eventually, the renaming attack will become obvious as more and more basic IP structures are standardized. Watermarking for copy detection will catch the IP illegally embedded inside of another IP. Finally, like constraint-based watermarking can never provide a certain authorship, any copy detection technique may miss some pirated IPs and catch some innocent users. However, the design of copy detection mechanism should have low false alarm rate as one of the key design objectives.

5. Summary

As we move into the information age, with the advances in the Internet and the World Wide Web, not only people have much easier access to the information they are seeking for, their privacy and intellectual property are becoming more vulnerable to attackers. In system design and VLSI CAD, there is also an urgent need for intellectual property protection techniques due to the reuse-based design methodology. This new design paradigm reuses existing IP blocks to build larger systems and thus greatly reduces the design cycle. However, it requires detailed information about the IP blocks. Designers of the IP blocks will not be willing to release such information unless their royalties are guaranteed. Therefore, the lack of effective protection schemes becomes a major barrier for the industrial adoption of design reuse to improve the design productivity.

The key challenge in IP protection is to keep IP's correct functionality. This is unique, compared to the state-of-the-art digital data watermarking and fingerprinting techniques as well as software protection and protocols for privacy protection over the Internet, which we will review in next chapter.

The constraint-based IP protection paradigm of *watermarking, fingerprinting, and copy detection* is the first set of self protection techniques for VLSI design IPs. We have seen, in this chapter, several examples and the overview of this approach. The rest of this book focuses on this IP protection paradigm with detailed discussion of its concepts, requirements, limitations, and applications.

For general interests in design reuse, we recommend readers the book "Reuse Methodology Manual" [84]. For a broader discussion about IP protection, we recommend the IP protection white paper released by VSI Alliance (editor: Ian R. Mackintosh). Part of the white paper is include in Appendix A and the whole document is available at VSI Alliance website *www.vsi.org*.

Notes

1 A recent survey [7] shows that half of the design projects have a 6-month time-to-market window and more than three quarters must be done within one year largely due to the emerging consumer products such as Internet appliances, set-top-boxes, wireless communications, and portable devices.

2 Sega develops and markets video entertainment systems, including the "Genesis" console and video game cartridges. Accolade is an independent developer and manufacturer of entertainment software, including game cartridges that are compatible with Genesis console as well as with other computer systems. Sega uses its own trademark security system (TMSS) to trigger a screen display of its trademark. Accolade reverse engineered Sega's video game programs to make its own game cartridges compatible with the Genesis console and copied the TMSS's initialization code. Sega sued Accolade for copyright and trademark infringement, and Accolade responded with the fair use defense to bar Sega from continuing to use its security system. (*http://laws.findlaw.com/9th/2/977/1510.html*).

3 Terabyte is a computer components broker which sells Intel math coprocessors to end-users. Terabyte did not purchase math coprocessors directly from Intel; rather it obtained the devices from other brokers and distributors. Intel sued Terabyte for redesigning slower math coprocessors and selling them as faster and more expensive math coprocessors. Intel tracked some of those "remarked" (by laser etching the particular model number on the chip) math coprocessors to Terabyte and found them either physically removed or covered and replaced with different markings bearing the Intel logo. (*http://laws.findlaw.com/9th/3/6/614.html*).

4 When Microsoft released Windows 1.0 with a similar graphical user interface (GUI) to Apple's. Apple complained and the two agreed to a license giving Microsoft the right to use and sublicense derivative works generated by Windows 1.0 in present and future products. When Windows 3.0 was released, Apple believed that it exceed the license, make Windows more "Mac-like," and infringe its copyright. (*http://laws.findlaw.com/9th/3/35/1435.html*).

5 Cadence and Avant! compete in the field of "place and route" software. Cadence sued Avant! for theft of its copyrighted and trade secret computer source code.(*http://laws.findlaw.com/9th/9715571.html*).

6 Connectix Corp. makes and sells a software program that enable buyers to play Sony PlayStation games on their computers instead of Sony PlayStation console. During the reverse engineering process, Connectix repeatedly copied Sony's copyrighted basic input-output system or BIOS, the software program that operates its PlayStation. (*http://laws.findlaw.com/9th/9915852.html*).

7 MP3 (MPEG-2 Group 3) is an audio file format that allows the file to be copied to computer hard disks or CDs. When compressed, MP3 files may be shared via the Internet, e-mail, and FTP. Napster's system enables its users to create MP3 music files and store them on individual computer hard drivers, to search for MP3 music files stored on other users' computer, and to transfer exact copies of the contents of other users' MP3 files from one computer to another via the Internet[139]. The Napster court case concluded that Napster had designed and operated a system that permits the infringing transmission and retention of sound recordings employing digital technology.

8 To enhance security and credibility, the meaningful signature message should first be encripted to pseudo-random bit stream, and then encoded as constraint, before they are embedded. This will be explained further in detail in Chapter 3.

9 In Figure 1.6, we encode each letter as a new edge between a pair of unconnected vertices. As one can imagine, the encoding table will have different meaning should we reorder the vertices.

Chapter 2

PROTECTION OF DATA AND PRIVACY

Although it is a new challenge to protect intellectual properties in VLSI designs, techniques and protocols for the protection of digital data, software, and privacy have been well-studied. The goal of this chapter is to survey the state-of-the-art protection techniques in these fields and analyze their applicability to VLSI design IP protection.

1. Network Security and Privacy Protection

The reuse-based design methodology forces designers to cooperate beyond their design team/company. The Internet and the WWW technologies help system designers to overcome the geographic barriers. Several Web-based distributed design environments have been proposed and demonstrated. For example, the WELD project in Berkeley targets a distributed collaborative EDA design system that is scaleable, adaptable, and secure. It includes a data manager, a server wrapper package, Java client package, proxy services, and a distributed workflow system[28]. Pia facilitates hardware/software co-design through geographically distributed co-simulation and integrates remotely located hardware into a co-simulation environment[73]. JavaCAD is based on a client-server architecture where clients are IP users and servers are IP providers. It provides an infrastructure for simulating and evaluating design over the Internet by remote method invocation[47]. Web-CAD is another tool for IP-based design analysis and simulation using the client-server architecture. It allows core vendors to make available very detailed core models without disclosing IP infomation[54]. A highly interactive universal client GUI is introduced in [24]. Combined with the concept of taskflow-oriented programming with distributed components, it creates a configurable computing environment for distributed networked design projects.

The use of the Internet and WWW enables the design, analysis, simulation, verification, and delivery of IP and IP-based system remotely. One important feature associated with these design environment is the security concern. However, since these approaches use standard client-server architecture and Java, such security concerns exist for all other network applications such as operating systems design, database design and management, networks and distributed systems, software execution and maintaining. Fortunately there have been plenty of studies and discussion on distributed network security and most of them are applicable to the Web-based VLSI CAD frameworks. Therefore, in the rest of this section, we briefly survey the security issues in networks.

Computer security consists of maintaining *confidentiality (or privacy, secrecy), integrity*, and *availability*. Modern (applied) cryptography tools and techniques play a very important role in providing security. These include: stream ciphers, block ciphers (the data encryption standard (DES), the fast data encipherment algorithm (FEAL), the international data encryption algorithm (IDEA), RC5, etc.), public key encryption systems (Rivest-Shamir-Adelman (RSA) encryption, Merkle-Hellman Knapsacks, El gammal, etc.), hash functions (manipulation detection codes (MDCs), message digest algorithms MD4 and MD5, message authentication code (MAC), etc.), digital signature algorithms, key establishment protocols and management techniques[112]. The potential threats in networks can be grouped into the following categories[127]:

- *Wiretapping*. Wiretap means to intercept communications, it can be done covertly such that neither the sender nor the receiver of a communication knows that the contents have been intercepted.
- *Impersonation*. Impersonation happens when a person gets other's authentication by guessing (the passwords), eavesdropping, avoidance (on weak or flawed authentication systems), or gets authentications that are well-known, trusted, or not existed[1].
- *Message confidentiality violations*. These violations, such as misdelivery and exposure, normally are human errors.
- *Message integrity violations*. Message integrity requires the message's correctness. Possible violations include: change the content, replace a message, change the source, redirect the message, destroy or delete the message.
- *Hacking*. Hackers usually develop tools to search widely and quickly for particular weaknesses and move swiftly and stealthily to exploit those weaknesses.
- *Code integrity violations*. Viruses, worms, Trojan horses, and other malicious code are designed to delete or replace running programs on a host and thus cause the code integrity problem.
- *Denial of service*. Connectivity failure, flooding, and routing problems are typical examples for denial of service.

The state-of-the-art techniques on network security controls include: encryption, access control, authentication, traffic control, firewalls, encrypting gateway, privacy enhanced e-mail, and so on.

Privacy protection is another problem related to IP protection. Privacy issues are exacerbating as the World Wide Web makes it easy for new data to be automatically collected and added to the database. Data entered into forms or contained in existing databases can be combined almost effortlessly with transaction records and an individual's every click. Internet service providers have the ability to keep track of the sites one visits and the software one downloads. Websites use cookies (bits of data that can be stored on PCs) to keep a record of visitors. This concern is increasing with the advances in data mining tools.

The Web cookie was invented by Lou Montulli for Netscape in 1994 to enable online shopping baskets. Before then, there was no way of figuring out what specific users did at websites, much less remembering what a customer ordered. Now there are "unfriendly" cookies such as stealth cookies hidden by third parties on Web pages (you visit a page and get tagged by cookies from sites you never visited) or security holes (Internet Explorer has one) that allow third parties to see your cookies.

One common type of carrier for cookies is the software known as E.T. application. This software plants itself in the depths of your hard drive and, from that convenient vantage point, starts digging up information. Often it is watching what you do on the Internet. Sometimes it is keeping track of whether you click on ads in software, even when you are not hooked up to the Internet[2].

E.T. applications take advantage of a simple fact: when we download software, most of us have no way of knowing what we are getting. More than 22 million people are believed to have downloaded the E.T. applications[37]. While the roots of E.T. applications go back to a program called Registration Wizard in Microsoft's Window 95[3], most of the current E.T. applications are embedded in shareware, the software that can be downloaded free from the Internet. For example, Conducent embeds ads in PKZip and CuteFTP[4]; Radiate places its software on Go!zilla and Free Solitaire[5]; zBubbles, a shopping tool by Alexa[6]; The browser from SurfMonkey[7]; The popular RealJukebox software RealNetworks[8].

The tools that kill cookies include: Cookie Monster which automatically delete some cookies as soon as they launch your hard drive; MacWasher from Webroot Software Inc. can be programmed to automatically wipe clean the web surfing history; and CookieCleaner which allows you to keep only the cookies you want and delete the rest.

Meanwhile, many research attempts also help Internet users surf the Web anonymously. Reiter and Rubin[137] discuss *Crowds*, an anonymity agent based on the idea that people can be anonymous when they blend into a crowd.

Rather than submitting HTTP requests through a single third-party, Crowds users submit their requests through the crowd, a group of Web surfers running the Crowds software. Goldschlag et al. [65] introduce *Onion Routing*, in which users submit encrypted HTTP requests using an onion (a layered data structure that specifies symmetric cryptographic algorithms). As data passes through each onion-router, one layer of encryption is removed and request arrives with only the IP address of the last onion-router. The Lucent Personalized Web Assistant[60] is used to insert pseudonyms into Web forms that request a user's name or e-mail address. It is designed to use the same pseudonyms consistently every time a particular user returns to the same site, but different at each site. TRUSTe[13] is a self-regulatory privacy initiative to build consumers' trust and confidence on the Internet. This online privacy seal program displays a privacy seal or "trustmark" on a home page informing visitors of the security practices conducted at the site.

2. Watermarking and Fingerprinting for Digital Data

Data watermarking, also known as data hiding, embeds data into digital media for the purpose of identification, annotation, and copyright. Recently, the proliferation of digitized media and the Internet revolution are creating a pressing need for copyright enforcement schemes to protect copyright ownership. Several techniques for data hiding in digital images, audios, videos, texts and multimedia data have been developed [16, 21, 43, 71, 128, 157, 166]. All these techniques take advantage of the limitation of human visual and auditory systems, and simply embed the signature to the digital data by introducing minute errors. The transparency of the signature relies on human's insensitiveness to these subtle changes.

Pfitzmann and Waidner [126] introduce and construct anonymous asymmetric fingerprinting schemes, where buyers can buy information anonymously, but can nevertheless be identified if they redistribute this information illegally. However, on finding a fingerprinted copy, the seller needs the help of a registration authority to identify the redistributer. Domingo-Ferrer[49] describes a construction for anonymous fingerprinting in which, on finding a fingerprinted copy, the seller needs no help to identify the dishonest buyer. In addition, the redistribution fraud can be proven to third parties.

In the rest of this section, we briefly survey the state-of-the-art protection techniques for text, image, audio, video, and other multimedia contents[9].

Text Documents

There are three major methods of embedding data into text documents: open space methods that encode through manipulation of white space, syntactic methods that utilize punctuation, and semantic methods that encode using manipulation of the words themselves.

Open space that has been used to hide data includes: space between vertical lines, space at the end of each line, space between words, baseline position of letters or punctuations, size and form of the letters or characters, and margin of the entire documents. For example, in line-shift coding, a bit 0 or 1 can be encoded as shifting a line vertically up or down slightly within a paragraph. This approach is based on the fact that most documents are formatted with uniform spacing between adjacent lines within a paragraph. Although the human eye is particularly adept at noticing deviations from uniformity, Low et al. [103] observe that vertical line displacements of $\frac{1}{300}$ inch and less, at 300 dot-per-inch resolution, can hardly be noticed by readers[10].

Syntactic method takes advantage of the ambiguity of punctuation and the circumstances when mispunctuation has low impact on the meaning of the text. For example, both phrases *bread, butter and milk* and *bread, butter, and milk* use commas correctly. We can use this alternation between forms to represent binary data. Other syntactic methods include the controlled use of contractions and abbreviations, and change of the diction and structure of text without significantly altering meaning or tone [14].

Semantic methods are similar to the syntactic method except that they change the words themselves instead of using the ambiguity of forms. More specific, synonyms (e.g., *big* and *large*, *small* and *little*, *smart* and *clever*) are assigned primary and secondary. Whenever there is place that both words can be used, we intentionally select the primary to embed 0 and the secondary for 1.

Both syntactic and semantic methods are robust against attacks like retyping or reformatting. However, human assistance is necessary to avoid changing the meaning of the text by the predetermined use of words and punctuations. In addition, their usage is limited by the nature of the methods themselves. Open space methods has numerous locations to embed information and all the techniques in this category can be automatic. The problem with this method is that all the data embedded will be removed by retyping the documents[11]. However, removing marks becomes difficult and requires human interaction as documents become rich and complicated.

Image

Information can be hidden into still images in many different ways, either directly in the spatial domain, or in a transformed domain such as the frequency domain. To hide information, direct message insertion may encode every bit of information into the image or selectively embed the message in "noisy" areas that draw less attention (e.g., areas where there is a great deal of natural color variation). The message may also be scattered randomly through the image. Redundant pattern encoding "wallpapers" the original image with the message. Common image watermarking approaches include: least significant bit insertion, masking and filtering, and algorithms and transformations.

Cox et al. [43] encode data as a sequence of independent and identically distributed Gaussian random variables and add them to the perceptually most significant DCT coefficients. By placing watermark in the perceptually relevant components of the original image, this technique provides a high level of robustness against many signal processing techniques aimed at eliminating noise from the image.

Koch and Zhao [93] describe a JPEG-based method for embedding label into images where the original image is divided into 8x8 blocks. A triple is chosen among the DCT coefficients at the middle frequencies in each block, and its components are modified to encode one bit.

Swanson et al. [158] present a watermarking method based on the addition in the frequency (DCT) domain of an spread spectrum signal. The signal is shaped by a perceptual mask that guarantees the invisibility of the hidden signal. The original image is segmented into blocks that are modified by single bits of the hidden message. The information decoding does not require the original image.

Bender et al. [14] propose the data-hiding scheme for image called *patchwork*. In this method, one bit is encoded by randomly choosing a certain number of pairs of pixels and modifying the difference in luminance level of each pair.

Audio, Video, DVD, and Other Multimedia Contents

Data hiding in audio signals tries to find the holes in human auditory system (HAS)[12]. For example, HAS has a fairly small differential range (e.g., loud sounds tend to mask out quiet sounds); it is unable to perceive absolute phase; and in most cases the common environmental distortions are ignored. Bender et al.[14] propose several techniques audio watermarking techniques: low-bit coding replaces the least significant bit of each sampling point by a coded binary string; phase coding substitutes the phase of an initial audio segment with a reference phase that represents the data; spread spectrum method encodes a stream of information by spreading the encoded data across as much of the frequency spectrum as possible; echo data hiding embeds data into a host audio signal by introducing an *echo*; other techniques include adaptive data attenuation, redundancy and error correction coding, and sound context analysis. Boney et al.[21] use a spread spectrum approach for audio watermarking. They filter a pseudo-noise (PN) sequence in several stages in order to exploit long-term and short-term masking effects of the HAS.

Video sequences consist of a series of consecutive and equally time-spaced still images. It is obvious that image watermarking techniques are directly applicable to video sequence. Hartung and Girod [71] employ a straightforward spread spectrum approach and embed an additive watermark into the compressed video. The watermarks are robust against standard signal processing and with a modified watermark detector against geometrical distortions like shift, zoom, and rotation. Swanson et al.[159] propose a multiscale watermarking method working on uncompressed video. The video is first segmented into scenes. Then a temporal wavelet transform is applied to each scene, yielding temporal low-pass and high-pass frames. The watermark is embedded into each of the temporal components of the temporal wavelet transform, and the watermarked coefficients are then inversely transformed to get the watermarked video. This scheme is robust against additive noise, MPEG compression, and frame drop.

The digital versatile disk (DVD) is the latest technology that has been developed to deliver data to the consumer. Protection has been a problem since the very beginning of DVD standard development[13] [141]. One way to secure the content on a DVD is to link a watermark verification process to the proper functioning of the DVD player. For instance, the player's output port would be enabled only upon verification of the watermark. Currently, there are several efforts in standardizing DVD copy protection technology. Most of them involve the use of watermarking and/or encryption techniques or other mechanisms including analog approaches for making images or video either not viewable or recordable[14].

Ohbuchi et al.[118] propose methods for embedding visible and invisible watermarks into 3-D polygonal models. Such models comprise of primitives like points, lines, polygons, and polyhedrons, which are attributed by their geometry and topology. They embed information by modifying vertices of pseudo-randomly selected triangles or tetrahedron from the mesh. Local variation of the mesh density can also be used to hide invisible watermarks.

Hartung et al.[70] watermark the facial definition parameters (FDP's) on MPEG-4 using a spread spectrum method. The watermarks are additively embedded into the animation parameters. Smoothing of the spread spectrum watermark by low-pass filtering and an adaptive amplitude attenuation prevents visible distortions of the animation. The watermark is not contained in the waveform representation of the depicted object, but in the semantics, i.e., the way the head and face move.

3. Software Protection

Software piracy has become a generic term for the illegal duplication of copyrighted computer software. General use of the term "piracy" encompasses three distinct categories of loss: (i) commercial piracy; (ii) corporate piracy; and (iii) softlifting. *Commercial piracy* refers to the illegal duplication of software for the purpose of distribution and sale. *Corporate piracy* typically takes the form of passing a piece of software around the office and placing it on multiple hard drives or copying onto a file server which is accessed by many people[15].

Softlifting occurs when a person copies a friend's software or brings a copy home from work for personal use[94].

The anti-piracy efforts mainly come from government and business organizations. The Software and Information Industry Association (SIIA *http://www.siia.net*) and the Business Software Alliance (BSA *http://www/nopiracy.com*) are the two primary business organizations fighting with software piracy. For years, they have been conducting studies on global software piracy, operating toll-free hotlines to encourage people to report corporate piracy and suspected incidents of software theft, providing free software and tips for software protection, among other fruitful efforts.

Generally, developers of computer software seek legal protection for intellectual property by using traditional legal mechanisms found in copyright, trade secret, patent, trademark, and licensing. Of these forms of protection, the most easily attainable protection is through copyright law, which makes it illegal to make or distribute copies of copyrighted material in the US without authorization[16]. State statutes and common law may be used to protect trade secrets embodied in computer software[17]. Patent law is designed to protect the idea behind an item, not merely the particular form in which the idea appears. A patent protects inventive advances in a technological process, a product, or a machine design. A trademark is a word, phrase, picture, symbol, shape, or other means the identifies the product's source. For many companies, the trademark is their most valuable asset and becomes a great marketing tool. There exist both federal and state trademark protection. The scope of protection provided by a license agreement often varies with the manner in which the respective software is marketed. License agreements emerged because of the perceived inadequacy of copyright and trade secret protection.

However, to seek these law protection, basic requirements have to be met and they cost money. More importantly, it needs time to process such applications. Furthermore, most companies are not interested in going to court. For example, among more than 200 lawsuits filed by the software publishers association (merged to SIIA on January 1, 1999), all but one were settled out of court. After the evidence of piracy has been discovered, they realize that litigation will serve no purpose. The only question remaining is how much money the company is willing to pay for its wrong-doing. Therefore, it is crucial how to collect sufficient evidence of infringement of convincing proof of authorship of the software. To this end, software developers have embedded company names, developer's signatures, and other marks into the software package in various methods. For example, place copyright statements in the source code as comments; use specific design style and/or variable naming convention to encode message; leave redundant or useless code segments in the final product; and so on. This is generally referred as "self-protection".

Software protection systems can be divided into two categories: *hardware based* and *software based*. In the former, the execution of the software is limited to the presence of specific devices such as CD-ROM, dongle, and smart card[9]. Licence number and keys have also been used to protect the software[6]. Recent progresses, noticeably the software obfuscation and watermarking techniques, have been reported in the second category. Obfuscation methods modify the compiled code to make decompilation harder, while watermarking approaches embed information into the codes and/or executables to prevent illegal reuse [38, 40, 154].

4. Summary

We have briefly reviewed the protection techniques for digital data, software, privacy, and network. The newly developed distributed collaborative EDA design systems, which leverage the Internet and WWW technologies, do have privacy and network security problems. However, most of these concerns are not unique to EDA and can be addressed by existing methods. On the other hand, state-of-the-art digital data watermarking and software protection techniques cannot be directly applied to the protection of VLSI CAD IPs because the IP's correct functionality must be maintained.

The protection techniques for digital data either use alternatives if they exist or introduce errors which cannot be detected by human. This eventually change the original digital data. So one cannot apply such techniques directly for the protection of IPs whose exact functionality need to be preserved. Fortunately, we experience that the implementation or structure of such IPs is that unique, i.e., there always exist large amount of solutions that guarantee the exact functionality. Therefore, we can apply constraint manipulation techniques to obtain a relatively unique solution rather than a random one and use the uniqueness of the obtained solution to protect our authorship. More specific, this conceptually new method, called *constraint-based watermarking*, translates the to-be-embedded signature into a set of additional constraints during the design and implementation of IP in order to uniquely encode the signature into the IP. The proof of authorship is shown by arguing the small probability for a random solution to satisfy all these extra constraints. The effectiveness of this generic scheme has been demonstrated at all stages of the design process [80].

Constraint manipulation is one of the widely used techniques in computer science and engineering. By carefully controlling (adding, deleting, modifying, etc.) the constraints, one can accomplish many tasks. In the context of testing and verification, constraints are added to check a desired property of a given circuit (c.f. Figure 3.11); in optimization algorithm developing, what the algorithm learns from recursively searching is expressed as new and/or modified constraints to help further search[107, 168]; in problem solving, one can add constraints to pursue solutions with particular structure; for problems whose exact solution is impossible or hard to find, constraints can be deleted (or relaxed) to determine a lower or upper bound, and can be added (or over-constrained) to get the other bound; $\cdots$. Recently, it finds a successful application in developing protection techniques.

Kahng et al. [80] first propose the generic approach of this idea as constraint-based watermarking and demonstrate how it can be used to protect the IPs in physical design [81]. Lac et al. [97, 98] show another application of embedding signatures and putting fingerprints in FPGA design. Later on, they improve the robustness of this protocol by hiding multiple small marks instead of one large global mark[99]. Then, Qu and Potkonjak [132, 134] build the necessary theoretical background for the constraint-based protection techniques. Besides, they also give the framework of how to compare different techniques quantitatively through the graph coloring problem. Qu et al. [133, 135] introduce the concepts of optimization-intensive and fair techniques to extend the applicable area from solely optimization-type problem to all kinds of problems, including decision problems like the Boolean satisfiability (SAT) problem. Such techniques also improve the quality of the embedded watermarks. Meanwhile, Kirovski et al. [89] watermark combinational logic synthesis solutions; Hong and Potkonjak [74, 75] propose techniques to protect DSP designs and design at the level of behavioral synthesis; Charbon [30] introduce the idea of hierarchical watermarking in IC design; Oliveira [120] develop robust techniques for watermarking sequential circuit designs; Khanna and Zane [85] show how to hide information in structure data by watermarking maps; Wolfe et al. [164] embed signatures in graph partitioning solutions; Caldwell et al. [27] use iterative techniques to fingerprint design IPs; Qu and Potkonjak [136] explain how to create different solutions instantaneously by constraint-addition; Kahng et al. [82] describe how to utilize the special structure caused by the additional constraints to develop fast pattern matching algorithms for copy detection; In [130, 131], data integrity techniques are combined with constraint manipulation to construct publicly detectable yet secure watermarks.

In the rest of the book, we discuss the key concepts of the *constraint-based IP protection paradigm* focusing on its three fundamental components: watermarking (in Chapter 3), fingerprinting (in Chapter 4) and copy detection (in Chapter 5). Watermarking has its goal to embed designer's digital signature into the design for later demonstration of authorship. Fingerprinting is a technique to deter people from illegally redistributing legally obtained IP by enabling the author of the IP to uniquely identify the original buyer of the resold copy. Copy detection is the mechanism to recover the embedded information. It is crucial for the entire IP protection process to quantify and qualify design similarity at an arbitrary level of design granularity among a set of suspicious code segments.

Notes

1 Eavesdropping happens when the authentication information is transferred and someone else is observing the communication; in a classic operating system flaw, the buffer for typed password has a fixed size, and overflow causes the OS to bypass password comparison and act as if a correct password is entered; well-known authentications refer to the cases when there exist accounts that do not require password or use a default password; trusted authentications are information of hosts or users that are trusted on other hosts, which are stored in the Unix *.rhosts*, *.login*, and *etchostsequiv* files.

2 They are called E.T. applications because after they have lodged in your computer and learned what they want to know, they do what Steven Spielberg's extraterrestrial did: phone home.

3 Registration Wizard lets purchasers dispense with snail mail and register their Window 95 software over the Internet. But it does something else too: it pokes around on the purchaser's hard drive, makes a list of other installed software and sends the information back to Microsoft.

4 PKZip is for compressing, storing and archiving files. CuteFTP is widely used by the MP3 crowd to fetch music files.

5 This E.T. software from Radiate, the advertising company formerly known as Aureate, has been embedded in 18 million people's computers and used their Internet connection to report back on what ads people were clicking on [37].

6 It monitors what users are doing online, even when they are not shopping, and reports back to Alexa.

7 SurfMonkey protects kids surfing the Web. It blocks questionable language and prevents children from accessing inappropriate Web pages. However, it requires a user ID and send home this information including phone number and e-mail address.

8 RealJukebox software lets users transfer music from the Net and their CDs to their hard drive so it can play in their computer. User's name and other identifying information are required for the software registration. Then whenever one puts a CD in the computer, his music choice and the machine's unique identifier are sent back to RealNetworks.

9 Most research work and products are on watermarking images. There is not much space in formatted text documents for watermarking; audio watermarking is more difficult than image and video watermarking due to the sensitivity of human auditory system; data can be hidden in video, as a sequence of images, by almost all the image watermarking techniques; watermarking for DVD, digital TV, 3-D polygonal models, and others share similar ideas.

10 For space at the end of line, data are encoded allowing for a predetermined number of spaces at the end of line[14]; word-shift coding shifts the location of a word horizontally (e.g. by $\frac{1}{150}$ inch or less [103]) within a text line to embed data; character modification method alters a particular feature of an individual character such as its height, position relative to other characters.

11 In addition, message will disappear with the end of line space in hard copy.

12 The human auditory system (HAS) operates over a wide dynamic range. It perceives over a range of power greater than one billion to one, and a range of frequencies greater than one thousand to one. Sensitivity to additive random noise is also acute. The perturbations in a sound file can be detected as low as one part in ten million[14].

13 Several media companies initially refused to provide DVD material until the copy-protection problem has been addressed.

14 The Data Hiding Subgroup (DHSG) of the Copy Protection Technical Working Group (CPTWG) has issued several calls for proposal in the area of data hiding and watermarking (*http:www.dvcc.comdhsg*). Digital Audiovisual Council (DAVIC) has also a special copyright issues group working on copy protection of images and video (*http:www.davic.org*)[165]

15 Corporate piracy rarely involves copying software for direct financial gain. However, a company will have purchased only one or a few copies of a program, yet dozens or hundreds of employees will be using the copies of that program.

16 The Copyright Act gives the author of copyrighted software five exclusive and separate rights to (i) reproduce the work; (ii) adapt or make derivative works; (iii) publicly distribute copies; (iv) publicly perform the work; and (v) display the copyrighted work.

17 A trade secret is any formula, pattern, device, or information used in the operation of a business to provide the business an advantage over competitors who do not know or use it[94].

Chapter 3

CONSTRAINT-BASED WATERMARKING FOR VLSI IP PROTECTION

We present the basic concepts of the *constraint-based watermarking technique*, which is designed to protect intellectual properties whose correct functionalities need to be preserved. We build the theoretical background for this generic approach and the framework to evaluate such techniques. We explain these by analyzing three watermarking techniques for the graph vertex coloring problem: the first one adds extra edges between some pairs of vertices and therefore forces them to be colored by different colors; the second one precolors a set of well-selected vertices according to the watermark; and the last one introduces new vertices and edges to the graph.

Since credibility (strength of the watermark) and overhead (performance degradation by the watermark) are the most important criteria for any efficient watermarking technique, we derive formulae that explicitly illustrate the tradeoff between high credibility and low overhead. For each of the above three GC watermarking techniques, we asymptotically prove that for almost all random graphs $G_{n,p}$, an arbitrarily high credibility can be achieved with the minimum 1-color overhead. Further watermarking features are analyzed based on numerical simulation on random graphs and experiments on graphs generated from real-life benchmarks.

The proposed constraint-based watermarking technique is not limited to optimization problems such as graph coloring. In this chapter, we also propose the first set of optimization-intensive watermarking techniques for decision problems. In particular, we demonstrate how one can select a subset of superimposed watermarking constraints so that the uniqueness of the signature and the likelihood of satisfying an instance of the satisfiability problem are simultaneously maximized. We have developed three SAT watermarking techniques: adding clauses, deleting literals, push-out and pull-back. Each technique targets different types of signature-induced constraint superimposition on an instance of the SAT problem. In addition to comprehensive experimental validation, we theoretically analyze the potential and limitation of the proposed watermarking techniques. Furthermore, we analyze the three proposed optimization-intensive watermarking SAT techniques in terms of their suitability for copy detection.

1. Challenges and the Generic Approach

As we have seen earlier, watermarking for the purpose of IP protection is difficult because it has to maintain the correct functionality of the initial IP. The *constraint-based watermarking technique* translates the to-be-embedded signature into a set of additional constraints during the design and implementation of IP in order to uniquely encode the signature into the IP. The proof of authorship is shown by arguing the small probability for a random solution to satisfy all these extra constraints.

1.1 Overview

Figure 3.1 outlines the general strategy for the constraint-based watermarking technique. It consists of two phases: watermark embedding and signature verification.

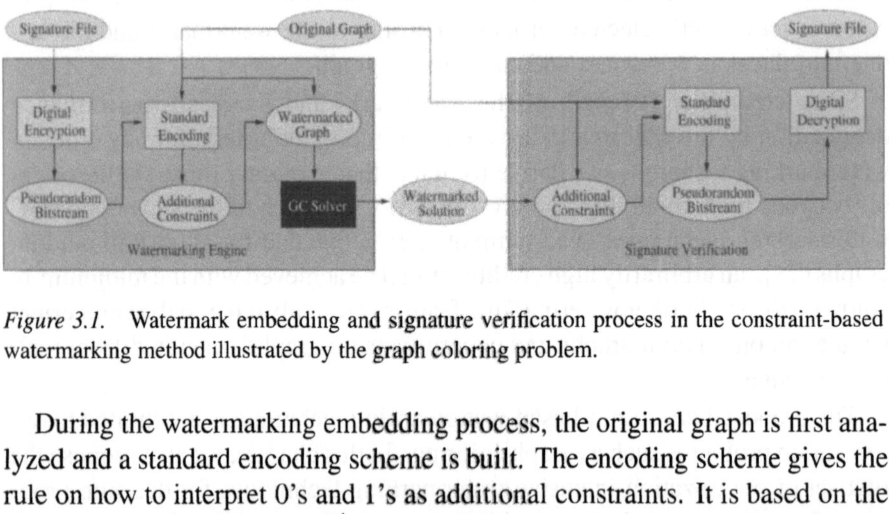

Figure 3.1. Watermark embedding and signature verification process in the constraint-based watermarking method illustrated by the graph coloring problem.

During the watermarking embedding process, the original graph is first analyzed and a standard encoding scheme is built. The encoding scheme gives the rule on how to interpret 0's and 1's as additional constraints. It is based on the property of the original graph[1] and independent of the author's signature file. Meanwhile, the author's signature is translated to a pseudorandom bitstream with the help of encryption and other cryptographic tools[2]. Then the standard encoding scheme takes this pseudorandom bitstream as input and outputs a set of additional constraints. These constraints are added into the original graph to form a watermarked graph. Finally the problem solver will be called to solve the watermarked graph (not the original one!). A watermarked solution is reported at the end of this phase.

To demonstrate the signature hidden in the found solution, the author has to identify the set of additional constraints (normally, certain special properties that a random found solution to the original graph does not necessarily have). The pseudorandom bitstream will then be retrieved from the standard encoding scheme. Using cryptographic tools again, one can decrypt this bitstream for the signature file.

1.2 Watermark Embedding Procedure

We have outlined the watermark embedding procedure. Our goal in this phase is to map the author's signature into additional constraints and enforce the problem solver to find a solution that satisfies these constraints. A pseudorandom bitstream is first generated based on the signature file and then encoded as constraints. Here we explain in detail two steps: the pseudorandom bitstream generation and the selection of constraints.

Suppose the signature file is in plain text, we first hash the message using a one-way hash function such as MD5 [138]. The hash result is then encrypted using our private key by an encryption system, for example RSA. Next, a stream cipher like RC4 is used to create the cryptographically strong pseudorandom bitstream. Note that up to this point, the generation of this pseudorandom bitstream is independent of the problem that we are solving. It is the encoding scheme, which connects the signature and the original problem, translates this sequence of pseudorandom 0/1's into constraints.

The selection of constraints to encode this bitstream can dramatically affect the strength of the watermark and the quality of the solution. A poor scheme will select constraints that either offer little proof of authorship, or cause innegligible degradation of solution's quality, or both. For example, we have discussed how to introduce extra edges to encode message in Chapter 1. An edge between two unconnected nodes that will most likely receive different colors[3] does not help us in building a credible watermark. However, an edge that increase the size of a large clique[4] will make the graph much harder to color and additional color may be needed consequently. Another concern for developing encoding scheme is to keep the watermarked problem similar to the original by keeping the properties such as graph's randomness and connectivity.

1.3 Signature Verification Procedure

To show the author's signature in the watermarked solution, we have to present two evidences: the existence of additional constraints and the correlation between these constraints and the claimed signature file.

The signature verification process consists of two steps. First, we create two (pesudorandom) bitstreams $\mathcal{S}_1$ and $\mathcal{S}_2$ of the same length and show that they are identical (or almost identical). $\mathcal{S}_1$ is obtained by mapping a set of selective additional constraints from the watermarked solution to 0/1's according to the inverse of a standard encoding scheme. S_2 is generated by a stream cipher (e.g., RC4) based on a (pesudorandom) seed that we choose. Note that these two "independently" created binary strings will differ in half of the bits on average. The event that S_1 and S_2 are identical, or almost identical, is highly unlikely. For two 128-bit strings, this happens with a probability of 2^{-128}, which is less than 10^{-37}. The occurrence of such a rare event reveals the correlation between the selected additional constraints and thus the watermarked solution and the seed that is used in the stream cipher.

Next, we must demonstrate that the seed is related to the our to-be-claimed signature file. To accomplish this, we decrypt the pesudorandom seed with our public key and show that the result is identical to the hash of our plain text signature file. Note that the hash result is also pesudorandom. For the same reason as we establish the correlation between watermarked solution and the seed, we conclude that the seed is indeed created from our signature because both the RSA system and the one-way hash function are hard to break.

1.4 Credibility of the Approach

By credibility of the approach, we mean how unique is the embedded watermark and whether the signature verification procedure is convincing. That is, can someone other than the author also "claim" the authorship to the IP? And if yes, how likely this may happen?

Numerous additional constraints could be easily identified from the (watermarked) solution. For example, in a solution to the graph coloring problem, any pair of vertices that are not connected by an edge but receive different colors can be viewed as a satisfied additional constraint. Because these two vertices do not have to be colored by different colors. This makes signature forgery a real possibility.

An adversary could fake a signature and then discover some additional constraints from the given IP (with other's watermark) such that these constraints coincide with the faked signaure in ASCII according to the standard encoding scheme. This might take some effort, but is not impossible, because adversary has full control of what signature he/she wants to forge and could fine tune this signature to match the selected set of constraints. However, once we include the cryptographic tools such as one-way hash function, stream cipher, and RSA encrption/decryption systems, into the watermarking process, it becomes extremely unlikely for adversary to obtain a successful forgery. The use of one-way hash function makes it computationally infeasible to find the plain text signature file which produces a given hash result. Therefore, the adversary cannot forge the signature based on his/her selected set of additional constraints. He/she could forge the signature first and then compute the set of corresponding constraints according to the standard encoding scheme. He/she will have a successful forgery if these additional constraints are satisfied by the given solution. But this is hard because the adversary can not change the solution at will to make this happen.

In sum, standard cryptographic tools and encoding schemes are not necessary for the constraint-based IP protection techniques, but they enhance the watermark's security. Particularly, as long as we believe these cryptographic systems are secure, we can claim that the author is the only one who can select a set of "random" additional constraints and go through the signature verification process to show these constraints are generated from the signature file.

1.5 Essence of Constraint Addition

The essence of constraint-based watermarking techniques is to add extra design constraints in order to get a rather unique solution. This is shown explicitly in Figure 3.2 for the GC problem.

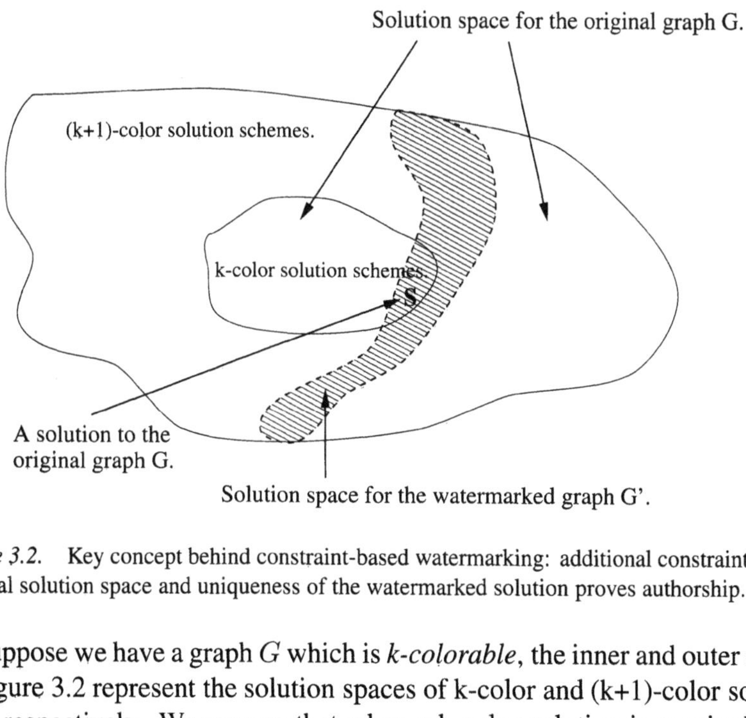

Figure 3.2. Key concept behind constraint-based watermarking: additional constraints cut the original solution space and uniqueness of the watermarked solution proves authorship.

Suppose we have a graph G which is *k-colorable*, the inner and outer regions in Figure 3.2 represent the solution spaces of k-color and (k+1)-color solutions to G respectively. We assume that when a k-color solution is required, every solution in the inner region has equal probability being picked. The shaded area is the solution space for the watermarked graph G', where we impose our signature as additional constraints. Since graph G' inherits all the constraints of graph G, a solution to G' is also valid for G. However, the solutions to G may violate the new constraints in G'. By coloring graph G' instead of G, we can obtain solutions to G and more important, we force the solutions fall into the shaded area. Denote n_k and n'_k the number of k-color solutions for graphs

G and G'. The chance to get a particular solution S from the constraints in G is $\frac{1}{n_k}$, which increases to $\frac{1}{n'_k}$ if from the more-constrained graph G'. When n_k is large and $n'_k << n_k$, the difference between $\frac{1}{n_k}$ and $\frac{1}{n'_k}$ is significant and becomes a credible evidence for the authorship.

High credibility depends not only on the amount of constraints, but also the "quality" of the constraints. For example, one constraint that cuts the solution space by half is definitely better than 20 constraints each cutting the solution space by less than 1%. Constraints for the GC problem are the edges: vertices connected by an edge have to receive different colors. One type of straightforward watermarks is extra edges. By translating signature as extra edges, we make the original graph more constrained, and some solutions to the original graph will become invalid for the watermarked graph. The solution space eventually shrinks. There are other interpretations of signatures as constraints. However, to have a transparent watermark, we require that the watermarked graph preserve the characteristics (e.g. connectivity, randomness, acyclicity.) of the original graph.

1.6 Context for Watermarking

As summarized by Kahng et al. [80], a generic watermarking procedure consists of the following components:

- An *optimization problem* with known difficult complexity. By difficult, we mean that either achieving an acceptable solution, of enumerating enough acceptable solutions, is prohibitively expensive. The solution space of the optimization problem should be large enough to accommodate a digital watermark.
- A well-defined *interpretation* of the solutions of the optimization problem as intellectual property.
- Existing *algorithms and/or off-the-shelf software* that solve the optimization problem. Typically, the "black box" software model is appropriate, and is moreover compatible with defining the watermarking procedure by composition with pre- and post-processing stages.
- *Protection requirements* that are largely similar to well-understood protection requirements for currency watermarking.

A non-intrusive watermarking procedure then applies to any given instance of the optimization problem, and can be attached to any specific algorithms solving it. Such a procedure can be described as:

- *A use model or protocols* for the watermarking procedure. In general, each watermarking scheme must be aware of attacks based on design symmetries, renaming, reordering, small perturbations (which may set requirements for the structure of the solution space), etc.

- Algorithmic descriptions of the *pre- and post-processing* steps of the watermarking procedure. Pre- and post processing preserve the algorithms and/or software as a "black box".
- *Strength and feasibility analyses* showing that the procedure satisfies given protection requirements on a given instance. Strength analysis requires metrics, and structural understanding of the solution space (e.g., "barriers" (with respect to local search) between acceptable solutions). Feasibility analysis requires measures of solution quality, whether a watermarked solution remains well-formed, etc.
- *General robustness analyses*, including discussion of susceptibility to typical attacks, discussion of possible new attacks, performance guarantees (including complexity analysis) and implementation feasibility.

1.7 Requirements for Effective Watermarks

In addition to maintaining the correct functionality of the IP, an effective watermark must satisfy the following properties:

- *high credibility:* The watermark should be readily detectable for the proof of the authorship. The probability of coincidence should be low.
- *low overhead:* The degradation of the software or design by embedding the watermark should be minimized.
- *resilience:* The watermark should be difficult or impossible to remove without the complete knowledge of the software or design.
- *transparency:* The addition of the watermark to software and designs should be transparent so that it can be used for existing design tools.
- *perceptual invisibility:* The watermark must be very difficult to detect. This is related to but not the same as the resilience problem.
- *part protection:* Ideally, a good watermark should be distributed all over the software or design in order to protect all parts of it.

In the following sections, we propose three watermarking techniques for the GC problems of random graphs, and investigate the impact of the corresponding watermarks to the solution space. In particular, the trade-off between credibility and overhead.

2. Mathematical Foundations for the Constraint-Based Watermarking Techniques

In this section, we lay out the mathematical foundation for the constraint-based watermarking approach by theoretically analyzing several watermarking techniques for the graph coloring (GC) problem. This also provides a framework for the evaluation and comparison of different watermarking methods.

2.1 Graph Coloring Problem and Random Graphs

We use the graph coloring (GC) problem as an example to illustrate our approach. The graph (vertex) coloring problem seeks to color a undirected graph with as few number of colors as possible, such that no two adjacent vertices receive the same color. It is formally defined as[61]:

Problem: Graph k-colorability
Instance: Graph $G(V, E)$, positive integer $k \leq |V|$
Question: Is G k-colorable, i.e., does there exist a function $f : V \rightarrow \{1, 2, ..., k\}$ such that $f(u) \neq f(v)$ whenever $(u, v) \in E$.

This problem is NP-complete and plays a very important role in complexity theory[61]. It also has numerous applications in various fields. For instance, Toft stated 75 interesting and easily-formulated graph coloring problems[115]. In VLSI CAD, the problems such as register allocation (as we have seen in the 4th order CF IIR filter example in Chapter 1), routing, cache-line coloring can all be easily induced from the GC problem. Many heuristics have been developed dedicated to it[174].

If we view the GC problem as a *constraint satisfying* problem, we want to minimize the number of colors subject to only one constraint: the two endpoints of any edge must receive different colors. The original problem instance, which is the graph itself, gives us all the constraints (i.e., edges) that any solution must meet. A watermarked solution to the GC problem is a coloring scheme that satisfies not only all these constraints, but also a set of additional constraints. These additional constraints are derived from authorship information and can be used for authenticate purpose.

The theory of random graphs was founded by Erdös and Rényi after Erdös had discovered, in the middle of this century, that probabilistic methods were often useful in tackling extremal problems in graph theory. The traditional way of estimating the proportion of graphs having a certain property is to obtain exact but complicated formulae. The new probabilistic approach is to approximate a variety of exact values by appropriate probability distributions and using probabilistic ideas.

The important discovery of Erdös and Rényi was that many important properties of graphs appear quite suddenly. If $\mathcal{Q}$ is a property, then for random graphs, either almost every graph has property $\mathcal{Q}$ or almost every graph fails to have property $\mathcal{Q}$. For example, let $M(n)$ be the number of edges in an n-vertex random graph, then if $\lim_{n\rightarrow\infty}(M(n) - \frac{1}{2}n\log n) = \infty$ almost all graphs are connected, and if $\lim_{n\rightarrow\infty}(M(n) - \frac{1}{2}n\log n) = -\infty$ almost all graphs are not connected. Bollobás' book "*Random Graphs*"[19] is the first systematic and extensive account of a substantial body of results from the theory of random graphs.

Random graphs play a very important role in many fields of computer science. The two most frequently occurring models of random graphs are $\mathcal{G}(n, M)$ and $\mathcal{G}(n, p)$. The first consists of all graphs with n vertices and M edges, the second consists of all graphs with n vertices and the edges are chosen independently with probability $p(0 < p < 1)$. We will focus on the second model and use these conventional notations: $G_{n,p}$ for an element of $\mathcal{G}(n, p)$, $q = 1 - p, b = \frac{1}{q}$, $\alpha(G_{n,p})$ is the independent number of graph $G_{n,p}$ (i.e., the maximal cardinality of independent sets.), and $chi(G_{n,p})$ denotes the chromatic number of $G_{n,p}$ (i.e., the minimum number of colors required to color the graph.). For almost all graphs $G_{n,p}$, we have [19]:

$$\alpha(G_{n,p}) = (2 + o(1)) \log_b n \tag{3.1}$$

$$\chi(G_{n,p}) = (\frac{1}{2} + o(1)) \frac{n}{\log_b n} \tag{3.2}$$

2.2 Watermarking Technique #1: Adding Edges

Technique Statement

Signature embedding:

Given a graph $G(V, E)$ and a message M to be embedded in G. Let $V = (v_0, v_1, \cdots, v_{n-1})$ and we encrypt the message into a binary string $M = m_0 m_1 \cdots$ (by stream ciphers, block ciphers, or cryptographic hash functions). Figure 3.3 shows how M is embedded into the graph G as additional constraints.

Input: a graph $G(V, E)$ and a message $M = m_0 m_1 \ldots$
Output: new graph with message M embedded

Algorithm:
copy $G(V, E)$ to $G'(V, E')$;
foreach bit m_i
{ find the nearest two vertices v_{i_1}, v_{i_2} that are not connected to vertex v_i;
if $m_i = 0$ add edge (v_i, v_{i_1}) to E'
else add edge (v_i, v_{i_2}) to E'
}
report graph $G'(V, E')$;

Figure 3.3. Pseudo code for technique # 1: adding edges.

By the nearest two vertices v_{i_1} and v_{i_2} which are not connected to vertex v_i, we mean that $i_2 > i_1 > i \pmod{n}$, the edges $(v_i, v_{i_1}), (v_i, v_{i_2}) \notin E$ and $(v_i, v_j) \in E$ for all $i < j < i_1, i_1 < j < i_2 \pmod{n}$. For example, in Figure 3.4, vertices 2 and 3 are the nearest two vertices that are not connected to vertex 0. The essence of this technique is to add an extra edge between two vertices, these two vertices have to be colored by different colors which may not be necessary in the original graph G. Figure 3.4 shows a graph of 11 nodes

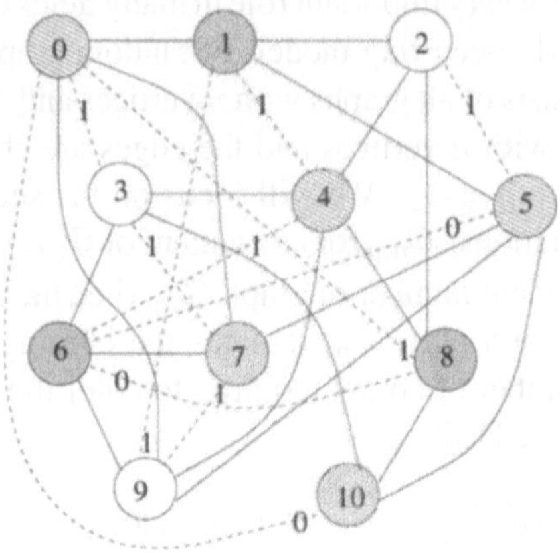

Figure 3.4. Example: a graph with message $1998_{10} = 11111001110_2$ embedded as additional edges.

with solid lines for original edges. The message $1998_{10} = 11111001110_2$ has been embedded by 11 dotted edges, each represents one bit marked on the edge. A 4-color scheme, $\{\{v_0, v_2, v_6\}, \{v_1, v_7, v_{10}\}, \{v_3, v_8, v_9\}, \{v_4, v_5\}\}$, is shown as well.

Signature recovering:

How can we read the watermark from the solution? Given the original graph, we claim that some pairs of vertices will have different colors. For example, in Figure 3.4, these pairs are $\{v_0, v_3\}, \{v_1, v_4\}, \{v_2, v_5\}, \cdots, \{v_{10}, v_0\}$. In the original graph, every such pair of vertices are not directly connected by an edge, so it is not necessary to assign them different colors. However we observe that this happens in the coloring scheme shown in Figure 3.4. For each such pair $\{v_i, v_j\}$, we can retrieve one bit of information by counting how many nodes in between (i.e., nodes with indices between i and j) are not connected to v_i. If there is none, the hidden bit is 0; if there is only 1, the hidden bit is 1; and if there are more than 1, reverse the order of v_i and v_j. This binary string is the (encrypted) message. In the same manner, it is not difficult to construct many other binary strings, even if the vertices have a standard order and the watermark is embedded in the well-accepted manner. For example, node 0 in Figure 3.4 has different color from both nodes 2 and 3, which are the nearest two vertices that are not connected to node 0. So both bits 0 and 1 can be claimed as the hidden bit in this case and one may have a different binary string. However, it will be hard to build one with a piece of meaningful information. In particular, if the original message is encrypted by one-way functions, forging a watermark with the same level of credibility needs to break the one-way functions.

Technique Analysis

The signature or message can be anything that is capable of identifying authorship. We can transfer it into binary (e.g., in ASCII), encrypt it by stream ciphers or cryptographic hash functions and assume the final bit stream is random. To have a quantitative analysis, we assume that exactly χ colors are required to color the graph $G_{n,p}$, where χ is given by[5]:

$$\chi(G_{n,p}) = \lceil \frac{n}{2\log_b n} \rceil \tag{3.3}$$

It follows immediately that after adding k extra edges into the graph $G_{n,p}$ according to the signature , the resulting graph remains random[6] with the same number of vertices and a new edge probability:

$$p' = p + \frac{2k}{n(n-1)} \tag{3.4}$$

So formula (3.3) for the chromatic number still holds, we denote this number by χ'. The *overhead* is defined to be $\chi' - \chi$, i.e., the number of extra colors required to color the watermarked graph. Intuitively, the more edges we add, the more colors we need to mark the graph. Since the number of colors is one of the most important criteria for the quality of coloring scheme, we want to keep this overhead as low as possible. One question is: how many edges can we add into the graph without introducing a large amount of overhead? Formally speaking: finding $k(n)$, the number of edges can be embedded into an n-vertex random graph, such that $\overline{\lim_{n\to\infty}} \chi' - \chi < \infty$.

Theorem 3.1
Adding $k(n)$ edges to a random graph $G_{n,p}$, for almost all $G_{n,p}$, $\overline{\lim_{n\to\infty}} \chi' - \chi = \infty$ iff $k(n) \in \omega(n \log n)$.

Proof:
In the original graph $G_{n,p}$, let $b = \frac{1}{1-p}$ and $\chi = \lceil \frac{n}{2\log_b n} \rceil$ as given by (3.3). After adding $k(n)$ extra edges, the edge probability increases to $p' = p + \frac{2k}{n(n-1)}$, and $b' = \frac{1}{1-p'}$, $\chi' = \lceil \frac{n}{2\log_{b'} n} \rceil$.

$$\chi' - \chi = \lceil \frac{n}{2\log_{b'} n} \rceil - \lceil \frac{n}{2\log_b n} \rceil \geq \frac{n}{2\log_{b'} n} - \frac{n}{2\log_b n} - 1 = \log_n (\frac{b'}{b})^{n/2} - 1$$

where $\frac{b'}{b} = \frac{1-p}{1-p'} = 1 + \frac{2k/n(n-1)}{(1-p)-2k/n(n-1)}$.

It is clear that $k(n) \in O(n^2)$, and further if $k(n) \in \Theta(n^2)$ then $\chi' - \chi \to \infty$ as $n \to \infty$. Therefore,

$$\lim_{n\to\infty}(\frac{b'}{b})^{\frac{n}{2}} = \lim_{n\to\infty}(1+\frac{2}{1-p}\frac{k}{n(n-1)})^{\frac{n}{2}} = \lim_{n\to\infty} e^{\frac{k}{(1-p)(n-1)}}$$

and

$$\overline{\lim_{n\to\infty}}\chi' - \chi \geq \lim_{n\to\infty} \log_n(\frac{b'}{b})^{n/2} - 1 = \lim_{n\to\infty} \frac{k}{(1-p)(n-1)\ln n} - 1$$

So, if $k(n) \in \omega(n \log n)$, $\overline{\lim_{n\to\infty}}\chi' - \chi = \infty$.

On the other hand, since

$$\chi' - \chi = \lceil\frac{n}{2\log_{b'} n}\rceil - \lceil\frac{n}{2\log_b n}\rceil \leq \frac{n}{2\log_{b'} n} + 1 - \frac{n}{2\log_b n} = \log_n(\frac{b'}{b})^{n/2} + 1$$

similarly, we can see if $k(n) \in O(n \log n)$, $\chi' - \chi$ will be bounded. □

Corollary 3.2 (1-color overhead)

Adding $k(n)$ edges to graph $G_{n,p}$, if $\lim_{n\to\infty}\frac{k(n)}{n\ln n} = l$, then $\overline{\lim_{n\to\infty}}\chi' - \chi \leq 1 + \frac{l}{1-p}$. In particular, if $k(n) \in o(n \log n)$, for almost all $G_{n,p}$, the overhead is at most 1.

A good watermark should be able to provide any desired level of confidence. That is, the authorship can be proved with a probability almost 1 when the graph goes large. Obviously one extra edge cannot bring high credibility. The following theorem answers the question: finding $k(n)$, the number of edges to be embedded into a n-vertex random graph, such that $Prob[\mathcal{E}] \to 0$ as $n \to \infty$, where $\mathcal{E}$ is the event that in a random solution all these $k(n)$ constraints are satisfied.

Theorem 3.3 (arbitrarily high credibility)

Adding $k(n)$ edges to a random graph $G_{n,p}$, let $\mathcal{E}$ be the event that a random solution to the original graph also satisfies all these $k(n)$ extra constraints. Then for almost all $G_{n,p}$, $\lim_{n\to\infty} Prob[\mathcal{E}] = 0$ if $k(n) \in \omega(\frac{n}{\log n})$.

Proof:

The event $\mathcal{E}$ is probabilistic equivalent to fixing a GC solution, then selecting $k(n)$ pairs of disconnected vertices and each pair do not have the same color. For random graph $G_{n,p}$, each vertex has $p(n-1)$ neighbors, and if the graph is colored by $\chi = \lceil\frac{n}{2\log_b n}\rceil$ colors as given by (3.3), in average there will be $\lfloor 2\log_b n\rfloor$ vertices for each color. Hence, when we select two disconnected vertices, the probability that they have different colors is $1 - \frac{\lfloor 2\log_b n\rfloor - 1}{q(n-1)}$. Assuming that $k(n)$ pairs of vertices are picked independently, then the probability that the vertices in each pair are of different colors is $(1 - \frac{\lfloor 2\log_b n\rfloor - 1}{q(n-1)})^{k(n)}$.

$$\lim_{n\to\infty} Prob[\mathcal{E}] = \lim_{n\to\infty}(1 - \frac{\lfloor 2\log_b n\rfloor - 1}{q(n-1)})^{k(n)}$$

$$\begin{aligned}
&\leq \lim_{n\to\infty}(1-\frac{\log_b n}{qn})^{k(n)} \\
&= \lim_{n\to\infty}(1-\frac{\log_b n}{qn})^{\frac{qn}{\log_b n}\frac{k(n)\log_b n}{qn}} \\
&= \lim_{n\to\infty} e^{-\frac{k(n)\log_b n}{qn}} \\
&= 0 \qquad\qquad (\text{ if } k(n)\in\omega(\frac{n}{\log n})=\omega(\chi).)
\end{aligned}$$

□

To summarize the "*adding edges*" technique, we conclude: adding $k(n) \in \omega(\frac{n}{\log n}) \cap o(n\log n)$ extra edges into graph $G_{n,p}$, as n goes large, arbitrarily high credibility can be achieved with at most 1-color-overhead. More precisely, we define the *watermark potential* (by adding edges) for graph $G_{n,p}$:

$$WP(G_{n,p}) = \chi(G_{n,p}) - \frac{n}{2\log_b n} \tag{3.5}$$

This function describes the power of the "*adding edges*" technique on random graphs. We list several properties of this function with respect to n (for p, similar results hold):

(a) $0 \leq WP(G_{n,p}) < 1$ for all graph $G_{n,p}$.

(b) *periodic:* χ is a non-decreasing step function and $\frac{n}{2\log_b n}$ is continuous and increasing. So $WP(G_{n,p})$ behaves periodically for different values of χ,

(c) *starting points:* χ increases by 1 at the start of each period $WP(G_{n,p})$ achieves its local maximum.

(d) *locally decreasing:* In each period, since χ is constant, as n increases, $WP(G_{n,p})$ decreases.

(e) *increasing period:* When n grows by 1, $\frac{n}{2\log_b n}$ will increase roughly by $\frac{1}{2\log_b n}$. Thus, the period is about $2\log_b n$. (a little larger than $2\log_b n$ to be more precise, since $b = \frac{1}{1-p}$ also increases.)

2.3 Watermarking Technique #2: Selecting MIS

Technique Statement

A maximal independent set (*MIS*) of a graph is a subset of vertices S such that vertices in S are not connected and vertices not in S are connected to at least one vertex of S. This second technique takes advantage of the fact that vertices in one MIS can all be labeled by a single color.

Signature embedding:

Given a graph $G(V, E)$ and a message M to be embedded in G. We order the vertices set $V = (v_0, v_1, \cdots, v_{n-1})$ and encrypt the message into a binary string $M = m_0m_1\cdots$. The message M is embedded into the graph G as shown in Figure 3.5. The idea is to select one or more MISs according to M, assign each MIS with one color and then color the rest of the graph. The MIS containing M is constructed in the following way: choose v_i as the first vertex of the MIS, where the binary expression of i coincides the first $\lfloor \log_2 n \rfloor$ bits of M; then we cut v_i and its neighbors from the graph since they cannot be in the same MIS as v_i; we reorder the vertices and select the next vertex of the MIS based on M. When we get a MIS, we color it with one color, remove it from the original graph and start constructing the second MIS if M has not been completely embedded.

A small example of an 11-node graph with the embedded message $1998_{10} = 11111001110_2$ is shown in Figure 3.6, where we use three colors to color the graph: $\{v_1, v_4, v_7, v_{10}\}, \{v_0, v_2, V_5, v_6\}$, and $\{v_3, v_8, v_9\}$. From 11 nodes, we choose node 7 to embed the first three bits of M ,111. Then all node 7's neighbors are crossed and the rest nodes are reordered; the node with the new index 3 is picked based on the next two bits 11; after cutting this node's neighbors, we obtain a MIS of the original nodes {1,4,7,10} which we mark by one color; reorder the rest 6 nodes and continue the procedure till M is completely embedded. Table 3.1 shows this procedure step by step.

```
Input: a graph G(V, E), a message M = m0m1 ...
Output: new graph with message M embedded
Algorithm:
current_graph = original_graph G(V, E);
previous_graph = current_graph;
MIS = φ;
do { if (current_graph is empty)
     { current_graph = previous_graph - MIS;
       previous_graph = current_graph;
       report MIS and reset MIS = φ; }
     if (current_graph has more than 2 vertices or is connected)
     { find the vertex v corresponding to the next ⌊log2 n⌋ bits of M;
       cut v and all its neighbors from the current graph;
       current_graph = current_graph - {v and its neighbors};
       MIS = MIS + v;
       reorder the vertices in current_graph;
       advance message M.
     } else  MIS = MIS + current_graph;
   }while(M is not empty)
report the current_graph;
```

Figure 3.5. Pseudo code for technique # 2: selecting MIS.

Signature recovering:

The selected MIS with a particular order of its vertices is the watermark. We can retrieve a binary string from this watermark by reconstructing the MIS in the specific order. For example, in Figure 3.6, 11111 is the information hidden

The purpose of introducing objective functions is to provide criteria that can be used to determine whether an additional constraint should be embedded or not during the optimization-intensive watermarking process. An objective function estimates the difficulty of determining the satisfiability of a formula. Obj_0 considers only the occurrence of l, l' and uses the ratio as a measure. Obj_1 takes into account the length of each clause that a literal and its complementary appears. In the second order objective function, not only l and l, but also their *neighbors* (the literals in the same clause) are considered. Therefore it provides more accurate estimation. Of course, better objective functions can be defined when we use more information from the SAT problem. Unfortunately, since the objective function will be called frequently, the computation cost of such function should be as low as possible. Usually, the accuracy of the objective function is at the expense of its complexity. For example, both Obj_0 and Obj_1 can be computed when the SAT instance is read in with the help additional storage. However, one more parse of the SAT instance is required to initialize Obj_2. A perfect objective function should be able to tell exactly the satisfiability of an instance and it cannot be computed in polynomial time unless $P = NP$.

For a given satisfiable formula, the optimization watermarking techniques do not guarantee the watermarked formula still satisfiable, but maximize this probability. Before we discuss the limitation of the proposed techniques, we mention a couple of properties of the defined objective functions:

$$
\begin{aligned}
Obj(\mathcal{F}) = 0 &\quad\Leftrightarrow\quad \mathcal{F} \text{ is } \textit{unsatisfiable} \\
Obj(\mathcal{F}) = \infty &\quad\Leftrightarrow\quad \mathcal{F} \text{ is trivially } \textit{satisfiable} \\
Obj(l) = \infty &\quad\Rightarrow\quad \mathcal{F} \text{ is satisfiable if } \mathcal{F} \text{ is satisfiable by assigning } l \text{ to be true}
\end{aligned}
$$

3.5.3 Limitations of the Optimization-Intensive Watermarking Techniques on Random SAT

The constant-probability SAT model

We adopt the model $J(n, r, p)$ for generating random SAT instances. A formula of this type consists of n clauses of r variables. A variable v is in the kth clause as an uncomplementary literal with probability p, as a complementary literal with probability p, and the kth clause does not contain variable v with probability $1 - 2p$.

Franco and Ho[58] proved that, for this model, almost all SAT instances can be solved in polynomial time if any of the following conditions holds:

$$p < \frac{0.4 \ln n}{r} \tag{3.11}$$

$$p > \frac{\ln n}{r} \tag{3.12}$$

following the same watermarking scheme, the hidden binary string becomes to 0010101 instead of 11111.

Technique Analysis

Our goal is to analyze this technique follow the framework we built in the previous section. In particular, we are interested in finding formulae for overhead and credibility.

First, we claim that after removing one randomly selected MIS, the remaining graph is still random with the same edge probability. One way to generate a random graph $G_{n+1,p}$ is to add one new vertex into a random graph $G_{n,p}$ and add an edge between the new vertex and each of the old vertex in $G_{n,p}$ with probability p. Reversing this procedure says that deleting one vertex from $G_{n,p}$ results in a random graph $G_{n-1,p}$. Since the neighbors of one vertex are also random, it follows that the graph will maintain its randomness after erasing one vertex and all its neighbors.

The first vertex of the MIS can be selected randomly, while the choices for the second vertex are restricted to $(1-p)n = qn$, because all the pn neighbors of the first vertex have been eliminated. In general, only $q^k n$ vertices are left as candidate for the (k+1)th vertex of the MIS. Therefore, we have:

Lemma 4.1
Given random graph $G_{n,p}$, almost all randomly selected MIS is of size $\log_b n$, where $b = \frac{1}{1-p}$.

The strength of the watermark relies on the uniqueness of the MISs we constructed as well as a specific order of the vertices in each MIS. To create a convincing watermark in a large graph, we have to add $\omega(\frac{n}{\log n})$ edges by the first technique. The same goal can be achieved by selecting only one MIS:

Theorem 4.2 (arbitrarily high credibility with 1-color overhead)
Given a random graph $G_{n,p}$, we select one MIS as in Figure 3.5. Let $\mathcal{E}$ be the event that in a random solution, all vertices in this MIS have the same color and they are in the order as specified by Figure 3.6. Then $\lim_{n\to\infty}$ *Prob* $[\mathcal{E}] = 0$. Furthermore, this introduces at most 1-color overhead.

Proof:

For a random graph $G_{n,p}$, the technique in Figure 3.6 gives us a MIS of size $\log_b n$ by Lemma 4.1. Given a fixed solution to $G_{n,p}$, event $\mathcal{E}$ has the same probability as: constructing all MISs of size $\log_b n$ with a specific order and one randomly picked MIS has all its $\log_b n$ vertices the same color. From the Stirling formula:

$$k! = (\frac{k}{e})^k \sqrt{2\pi k} e^{\alpha_k}$$

where $\frac{1}{12k+1} < \alpha_k < \frac{1}{12k}$, we have:

$$
\begin{aligned}
Prob[\mathcal{E}] &= \frac{1}{\frac{(nq)\cdot(nq^2)\cdot\dots\cdot(nq^{\log_b n-1})}{(\log_b n)!}} = (\frac{1}{q})^{\frac{\log_b n(\log_b n-1)}{2}}(\frac{1}{n})^{\log_b n-1}(\log_b n)! \\
&= (b^{\log_b n})^{\frac{\log_b n-1}{2}}(\frac{1}{n})^{\log_b n-1}(\log_b n)! = n^{\frac{1-\log_b n}{2}}(\log_b n)! \\
&= n^{\frac{1-\log_b n}{2}}(\frac{\log_b n}{e})^{\log_b n}\sqrt{2\pi\log_b n}e^{\alpha_{\log_b n}} \\
&= \sqrt{2\pi n\log_b n}e^{\alpha_{\log_b n}}(\frac{\log_b n}{e\sqrt{n}})^{\log_b n} \\
&\to 0 \quad (n \to \infty)
\end{aligned}
$$

where $\frac{1}{12\log_b n+1} < \alpha_{\log_b n} < \frac{1}{12\log_b n}$.

It costs exactly one color for the selected MIS, and coloring the remaining graph requires no more than the number of colors for the original graph. Therefore, this introduces at most one extra color overhead. □

By selecting one vertex from an n-vertex graph, we can embed $\lfloor \log_2 n \rfloor$ bits. From Lemma 4.1, at most $\log_2 \cdot n \log_b n$ bits of information could be embedded into the MIS. To embed long messages, we have to construct more MISs,[8] which may result in huge overhead.

Theorem 4.3
Given a random graph $G_{n,p}$, if we select $k(n)$ MISs as in Figure 3.5, assign each MIS one color and color the rest of the graph, then the overhead is at most $k(n)$ and on average at least $\frac{k(n)}{2}$.
Proof:

The first part is trivial from the fact that χ is non-decreasing in terms of n.

By Lemma 4.1, the MIS is of size $\log_b n$. Assuming the message is random, after we cut this MIS from the original graph $G_{n,p}$, the remaining graph will still be random with $n' = n - \log_b n$ vertices and the same edge probability p. Therefore, from formula (3.4), we need $\chi' = \lceil \frac{n'}{2\log_b n'} \rceil$ colors to color this remaining graph, taking into account one more color for the selected MIS, we use a total of $\chi' + 1$ colors to color the original graph $G_{n,p}$.

$$
\begin{aligned}
\chi' + 1 - \chi &= \lceil \frac{n'}{2\log_b n'} \rceil + 1 - \lceil \frac{n}{2\log_b n} \rceil \\
&= \lceil \frac{n - \log_b n}{2\log_b (n - \log_b n)} \rceil + 1 - \lceil \frac{n}{2\log_b n} \rceil \\
&= \lceil (\frac{n}{2\log_b n} - \frac{1}{2})\log_{(n-\log_b n)} n \rceil + 1 - \lceil \frac{n}{2\log_b n} \rceil \\
&\geq \lceil \frac{n}{2\log_b n} - \frac{1}{2} \rceil + 1 - \lceil \frac{n}{2\log_b n} \rceil \\
\text{since} \quad & \log_{(n-\log_b n)} n > 1
\end{aligned}
$$

For a uniformly distributed real number x, $Prob(\lceil x - \frac{1}{2} \rceil = \lceil x \rceil) = 0.5$. Therefore, when we construct one MIS by Figure 3.5, we will introduce one extra color overhead with probability at least 50%. And when we construct two MISs, for sure we will introduce at least one-color-overhead since $\lceil \chi - 1 \rceil = \lceil \chi \rceil - 1$. In general, when $k(n)$ MISs are selected, the size of the remaining graph $n' \geq n - k \log_b n$ because the size of MIS decreases with the size of the graph. So

$$\chi' + k(n) - \chi \geq \lceil \frac{n}{2 \log_b n} - \frac{k(n)}{2} \rceil + 1 - \lceil \frac{n}{2 \log_b n} \rceil \sim \frac{k(n)}{2}$$

□

2.4 Watermarking Technique #3: Adding New Vertices and Edges

Technique Statement

Signature embedding:

Given a random graph $G_{n,p}$ and a message M to be embedded. We order the vertices set $V = (v_0, v_1, \cdots, v_{n-1})$ and encrypt the message into a binary string $M = m_0 m_1 \cdots$ which is then embedded into $G_{n,p}$ as follows: introduce a new node v, take the first $\lfloor \log n \rfloor$ bits from M, find the corresponding vertex v' and connect it to v; take the next $\lfloor \log n - 1 \rfloor$ bits and locate the next vertex to which v is connected ($n - 1$ since v' has to be excluded); continue till we add np edges starting from v and get a new graph $G_{n+1,p}$; introduce another new node if M has not been completely embedded. We color the new graph, restrict the coloring scheme to the original graph $G_{n,p}$ and we have a solution with message M embedded.

Signature recovering:

This watermark is hard to detect because of the invisibility of the new added nodes and their associated edges. To exhibit the hidden signature in a colored graph, we have to go through the signature embedding procedure again and show that the encrypted signature can be added into the colored graph as edges to the newly inserted vertices without any conflicts. This has to be coupled with a statement of the unlikelihood that this happens for any random message. As we discussed earlier, many different binary strings can be generated in the same way from the same colored graph, but to fake one corresponds to a one-way function with a specific information is not easy.

Technique Analysis

Suppose k new nodes have been added into the initial graph $G_{n,p}$ to accommodate the message, similar to the previous two techniques, it is clear that the embedded graph is an instance of $G_{n+k,p}$[9]. This guarantees that randomness of the watermarked graph and hence the validity of the formula (3) which implies an overhead in the amount of $\chi' - \chi = \lceil \frac{n+k}{2\log_b(n+k)} \rceil - \lceil \frac{n}{2\log_b n} \rceil$, where $b = \frac{1}{1-p}$.

We have defined the watermark potential for graph $G_{n,p}$ as $WP(G_{n,p}) = \chi(G_{n,p}) - \frac{n}{2\log_b n}$. A large $WP(G_{n,p})$ means there is still room for adding new nodes and/or edges into $G_{n,p}$ without introducing a new color, especially at the starting point of each period (property (c) of function $WP(G_{n,p})$ in section 3.3.2). From the step function nature of χ, we have

Theorem 5.1 (1-color overhead)
Given a random graph $G_{n,p}$, we introduce $k(n)$ new vertices and associate edges based on the signature, then for almost all $G_{n,p}$, the overhead is at most 1 if $k(n) \in o(\log n)$.

A graph of χ colors is essentially a partition of the vertices to χ independent sets. The neighbors of any new vertex can be selected randomly from these χ set. However, to add one new vertex v without bringing a new color, v's neighbors have to be chosen from at most $\chi - 1$ independent sets. It is not hard to see that when many edges have to be added, it is unlikely that none of these edges ending into a specific independent set.

Theorem 5.2 (arbitrarily high credibility)
We build graph $G_{n+1,p}$ from a given random graph $G_{n,p}$ by introducing one new vertex and np edges. A coloring scheme to the initial $G_{n,p}$ is obtained by coloring $G_{n+1,p}$. Let $\mathcal{E}$ be the event: add a vertex v to the colored graph $G_{n,p}$, connect v to np random vertices, and v does not require a new color. Then $\lim_{n\to\infty} Prob[\mathcal{E}] = 0$ for almost all $G_{n,p}$.

2.5 Simulation and Experimental Results

2.5.1 Numerical Simulation for Techniques # 1 and # 2

We conduct simulation in the ideal case assuming we know how to color the graph optimally. In the "adding edges" technique, we add extra edges into the original graph corresponding to the signature. Figure 3.7 shows for graph $G_{n,0.5}$ ($500 \le n \le 1000$), the number of edges can be added (y-axis) with 0-overhead ($k_0(n)$, shown as black dots) and 1-color overhead ($k_1(n)$, shown as gray triangles), the curve in between is the difference of $k_0(n)$ and $k_1(n)$. Revisiting the properties of the watermark potential function, we see that $WP(G_{n,p})$ describes correctly the amount of information can be embedded into graph $G_{n,p}$.

In Figure 3.8, for graph $G_{n,0.5}$ ($500 \le n \le 550$),the numbers of MISs (y-axis) that can be constructed within 2-color overhead are given. One observation is that the number of MISs as a function of n for the same number of overhead is piecewise constant. This has been predicted from the proof of Theorem 4.3.

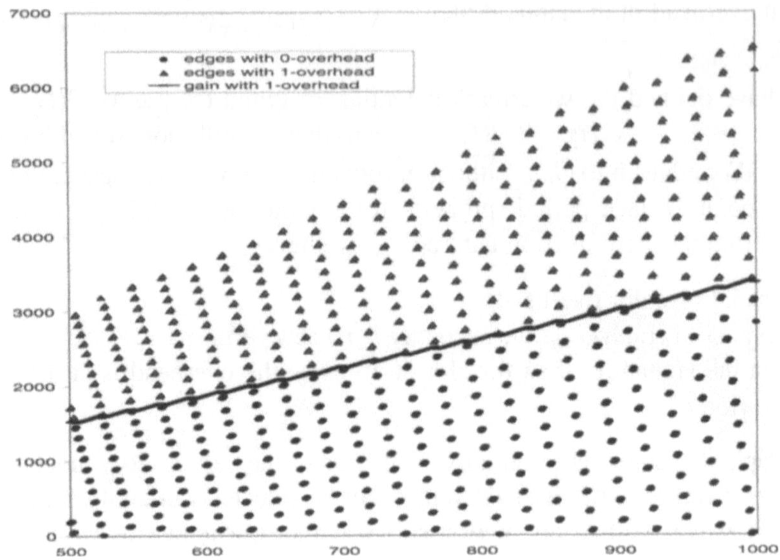

Figure 3.7. Numerical simulation data for technique # 1: the number of edges can be added in with 0- and 1-color overhead for random graph $G_{n,0.5}$ ($500 \leq n \leq 1000$). The curve in between shows the gain (in terms of the number of extra edges) with one extra color.

Another fact is that when we select one MIS, with 50% probability there will be a 1-color overhead. The reason is that the increment on $\frac{n}{2\log_b n}$ by selecting one MIS is around $\frac{1}{2}$ and *Prob*$[\lceil x - \frac{1}{2} \rceil = \lceil x \rceil] = 0.5$.

2.5.2 Experimental Results

The main goal of the experiment is to compare the difficulty of coloring the original graphs vs. the watermarked graphs, as well as the quality of the solution. For this purpose, we choose three types of graphs: random graphs $G_{n,p}$, graphs generated from real-life benchmarks, and the DIMACS challenge graph. For each type of graphs, we do the simulation in three steps: (1) color the original graph, (2) apply the watermarking techniques to embed a random message, (3) color the watermarked graph. Each graph is colored 10 times and the average result is reported. All experiments are conducted on 200MHz UltraSparcII and 40 MHz SPARC 4 processors using the algorithm in [88]. The same parameters are used for the original and watermarked graph.

Table 3.2 shows the results on random graphs $G_{n,0.5}$, and the corresponding watermarked graphs by adding n and $2n$ random edges or by selecting one MIS. The columns labeled *color* are the average numbers of colors on 10 trials for each instance, while the *best* columns are the best solutions from the 10

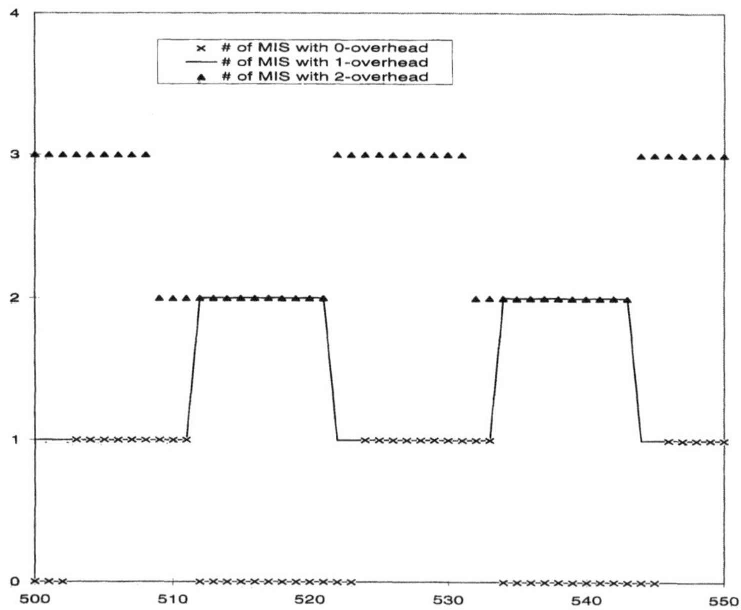

Figure 3.8. Numerical simulation data for technique # 2: the number of MISs that can be selected to embed signature with 0-, 1-, and 2-color overhead for graph $G_{n,0.5}$ ($500 \leq n \leq 550$).

Original $G_{n,0.5}$			Adding n Edges			Adding $2n$ Edges			Selecting 1 MIS		
n	color	best	color	best	mesg	color	best	mesg	color	best	mesg
125	19	19	19	19	125	19	19	250	19	19	42
250	30	30	30.2	30	250	30.2	30	500	30	30	80
500	50.1	50	50.4	50	500	50.6	50	1000	50.2	50	81
1000	85.8	85	86	85	1000	86.8	86	2000	86	85	110

Table 3.2. Coloring the watermarked random graph $G_{n,0.5}$: (i) adding n edges; (ii) adding $2n$ edges; (iii) selecting one MIS ($n = 125, 250, 500, 1000$).

trials, and the columns *mesg* measure the amount of information (in bits) being embedded in the graph.

Table 3.3 is the result on dense/sparse random graphs. For dense graphs $G_{n,0.9}$, there is not much space left to add extra edges, so it is expensive to watermark dense graphs by adding edges. On the other hand, the size of MIS for dense graph is relatively small, therefore very limited information can be embedded by selecting MISs. For sparse graphs $G_{n,0.1}$, both techniques perform well.

Original $G_{n,p}$			Add n	Add $2n$	Select	Select
n	p	color	Edges	Edges	1 MIS	2 MISs
125	0.9	46	49.1	52	47	47.3
250	0.9	77.9	79	82	77.2	77
250	0.1	9	9.1	9.8	9	9.8
500	0.1	13.9	14	14.2	14	14.2

Table 3.3. Coloring the watermarked dense/sparse graph $G_{n,p}$ for $n = 125, 250, 500$ and $p = 0.1, 0.9$.

When applying to the on-line challenge graph at the DIMACS site [174], for the graph with 1000 vertices and 249826 edges which implies an edge probability slightly larger than 0.5, we restrict the run-time to 1 hour and get the results from 10 trials shown in Table 3.4. In the 10 trials for the original graph, we find two 85-color solutions and the average number of colors is 86.1. The second column is the amount of information (in bits) being added into the graph. The last column shows the probability of coincidence, where low coincidence means high credibility. One can see both methods provide high credibility with little degradation of the solution's quality.

Edges Added	Information	Colors	Overhead	Coincidence
0	0	86.1	-	-
500	500	85.8	-0.3	2.89e-03
1000	1000	87	0.9	9.55e-06
2000	2000	87	0	9.12e-11
3000	3000	87	0.9	8.71e-16
MIS Selected	Information	Colors	Overhead	Coincidence
1	110	86.2	0.1	3.52e-17
2	210	86.4	0.3	1.24e-33
3	300	87	0.9	8.51e-50
4	390	87.4	1.3	5.85e-66
5	480	87.4	1.3	2.06e-82
6	590	87.9	1.8	7.24e–99

Table 3.4. Coloring the watermarked DIMACS benchmark.

For the technique of "adding new vertices and edges", we start from a random graph $G_{125,0.5}$ and introduce new vertex (and certain number of edges to keep the edge probability) one by one till we reach an instance of $G_{549,0.5}$. Then we color each of these 425 graphs 10 times and plot the average number of required colors in Figure 3.9. The results for the last 50 instances are enlarged as shown in Figure 3.10

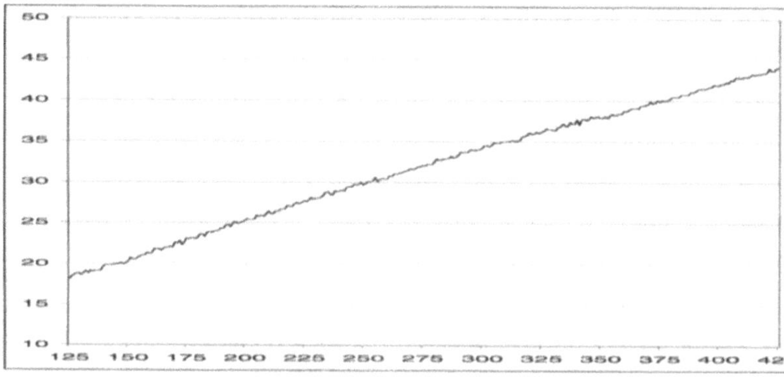

Figure 3.9. Coloring the watermarked graph $G_{n,0.5}$ by technique # 3: adding new vertex (and its corresponding edges) one by one for $n \in [125, 549]$.

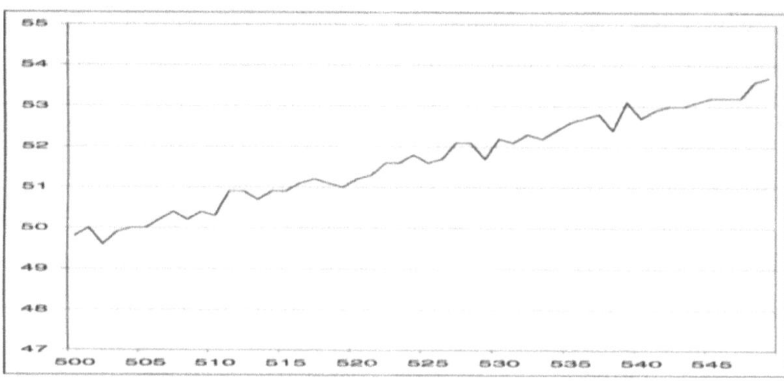

Figure 3.10. The last 50 instances of graph $G_{n,0.5}$ in Figure 3.9 ($500 \leq n \leq 549$).

The graph coloring problem has a lot of applications in real life, for example, the register allocation problem, the cache-line coloring problem, wavelength assignment in optical networks, and channel assignment in cellular systems. The instances of GC problems based on register allocation of variables in real codes and the optimal solutions are available at [175]. We watermark these graphs and then color them. The *fpsol2* and *inithx* instances are colored in 1 $\sim$ 3 minutes, while the others are all colored in less than 0.5 minute. Table 3.5 reports the details. The first four columns shows the characteristic of the original graph and the known optimal solution; the next two are for technique #1, showing the number of edges (information in bits) being embedded and the overhead; followed by two columns for technique #2, where the *Size* columns are the number of vertices in the selected MISs. The last two columns are for technique #3, where we compute the average edge probability of the original graph and add edges to keep this probability unchanged. Again, in almost all examples, there is no overhead.

Original Instance				Add n Edge		Select 1 MIS		Add 1 Vertex	
Instance	|V|	|E|	k	Edges	Colors	Size	Colors	Edge	Colors
fpsol2.i.1.col	496	11654	65	496	*65*	229	*65*	47	*65*
fpsol2.i.2.col	451	8691	30	451	*30*	90	*30*	38	*30*
fpsol2.i.3.col	425	8688	30	425	*30*	64	*30*	40	*30*
inithx.i.1.col	864	18707	54	864	*54*	347	*54*	43	*54*
inithx.i.2.col	645	13979	31	645	*31*	89	*31*	43	*31*
inithx.i.3.col	621	13969	31	621	*31*	64	*31*	45	*31*
mulsol.i.1.col	197	3925	49	197	*49*	61	*49*	40	*49*
mulsol.i.2.col	188	3885	31	188	32	17	32	41	*31*
mulsol.i.3.col	184	3916	31	184	32	12	32	42	*31*
mulsol.i.4.col	185	3946	31	185	*31*	12	*31*	42	*31*
mulsol.i.5.col	186	3973	31	186	*31*	12	*31*	42	*31*
zeroin.i.1.col	211	4100	49	211	*49*	87	*49*	39	*49*
zeroin.i.2.col	211	3541	30	211	*30*	56	*30*	33	*30*
zeroin.i.3.col	206	3540	30	206	*30*	51	*30*	34	*30*

Table 3.5. Coloring the watermarked real-life graphs by: (i) adding n edges; (ii) selecting one MIS; (iii) adding one new vertex. |V|: number of vertices; |E|: number of edges; k: minimal number of colors.

3. Optimization-Intensive Watermarking Techniques

We have explained the generic approach of the constraint-based watermarking techniques for optimization related IP protections. We now extend it to the decision problems represented by the Boolean Satisfiability (SAT) problem.

3.1 Motivation

The proposed constraint-based watermarking technique is conceptually different from those designed for data hiding in artifacts (digital images, audio, video, text, and multimedia). This technique is applicable to protect IPs that can be properly mapped to an optimization problem such as graph coloring. In the watermarking process, a (digital) signature is translated and then embedded into the original optimization problem as additional constraints. It is this watermarked problem that will be solved and the solution remains valid for the initial problem since all original constraints are met. The authorship is provided by showing that a randomly selected solution to the initial problem can rarely survive all the signature-based extra constraints.

However, there are two factors that limit the usage of this generic technique. First, the embedding of watermarks can make a problem over-constrained and we then have to consider the quality of the watermarked solution. Although both theoretical and experimental results [132, 88] suggest the degradation of the solution's quality is negligible, it remains as one of the biggest concerns for IP providers to watermark their IPs. Secondly, this technique cannot be used directly to watermark decision problems because of the natural difference between optimization and decision problems. For decision problems, not only the degradation of solution's quality, but also the solution itself become a problem. For example, if a watermarked satisfiability problem is not satisfiable, then we have to ask ourselves whether the problem instance itself is unsatisfiable or our watermark makes it unsatisfiable.

On one hand, it may not be hard to find a solution to an optimization problem. What makes it difficult and interesting is to find an *optimal solution.* In most cases, sacrificing the solution's quality for proof of authorship may not be acceptable. On the other hand, decision problems, represented by the Boolean satisfiability (SAT) problem, play the central role in theoretic computer science and find numerous applications in various fields. SAT is the first computational task shown to be *NP-hard* by Cook (1971). Due to its discrete nature, SAT appears in many contexts in the field of VLSI CAD, such as automatic pattern generation, logic verification, timing analysis, delay fault testing and channel routing. Therefore, we need new and more powerful watermarking techniques to improve the quality of the solutions to the optimization problems and to protect the decision problems.

A Motivational SAT Example

Consider the formula $\mathcal{F}$ of 13 variables in the standard conjunctive normal form (CNF) [173] that we have shown in Chapter 1:

$$(x_3+x_4)(x_1'+x_3')(x_9'+x_8')(x_3+x_{11})(x_{13}'+x_{12}')(x_1+x_2+x_3')(x_6+x_8+x_9)$$
$$(x_1+x_3+x_7)(x_4+x_{11}+x_{10})(x_1'+x_2'+x_3')(x_{12}+x_5'+x_2)(x_7'+x_6+x_4+x_5)$$
$$x_2\cdot x_5\cdot(x_2'+x_3+x_5+x_9)(x_3+x_5+x_7'+x_{12})(x_6+x_4'+x_8+x_{11}+x_{13}')$$

We encode a message into new clauses by mapping letters "a - z" to $x_1, x_1', \cdots, x_{13}, x_{13}'$ alphabetically. We have showed that the phrase "*A red dog is chasing the cat*" will be translated to seven extra clauses: $x_1(x_9'+x_3+x_2')(x_2'+x_8+x_4)(x_5+x_{10})(x_2+x_4'+x_1+x_{10}+x_5+x_7'+x_4)(x_{10}'+x_4'+x_3)(x_2+x_1+x_{10}')$. And after embedding these clauses to formula $\mathcal{F}$, only *12* of the previous 256 truth assignments remain valid. The authorship of the found solution, one of these 12, is claimed probabilistically. Assuming that the SAT solver has equal probability of finding any specific solution. We argue that one has a chance of $\frac{1}{256} \approx 0.39\%$ to get it from the original formula, while this chance increases to $\frac{1}{12} > 8.3\%$ if one solves the watermarked formula.

The problem arises if we use the same technique to embed *"A red dog is chasing the bee"*. None of the previous 256 truth assignments can satisfy the clauses based on this message and we will see the problem *unsatisfiable*! Remember that the original formula **IS** satisfiable, the purpose for adding constraints is to protect the solution we find. There is no need to protect a solution that is incorrect and useless. This happens because the extra constraints may over-constrain the problem. For optimization problem, such over-constraineness is less visible since we will (almost) always find a solution, just the quality of the solution matters. For decision problem, the entire solution could be changed. The question we are facing now is how to add constraints such that the watermarked problem is not over-constrained while still provides a sufficient proof of authorship.

Solution: Optimization-Intensive Techniques

In this section, we discuss the *optimization-intensive techniques* that solve the above problem. The basic idea is to embed the message in an "optimal" way such that the probability of changing the solution to the decision problem (or degrading the quality of an optimization problem's solution) is minimized.

Recall that in Figure 3.1, the encryption of the signature file and the development of the standard encoding scheme are independent[10]. When we convert the pseudorandom bitstream generated from the signature file to constraints, we do a "blind encoding", which means that every bit will be translated into constraints. However, this is not necessary if we take a close look at how the authorship is proved. We show it in the probabilistic way by arguing that getting such a particular solution accidentally is very unlikely, which means the authorship can never be certain. Therefore, as long as we can give a convincing proof, we do not have to embed the entire signature file.

In the new optimization techniques, we replace such "blind encoding" with *selective encoding*. In particular, before we embed any watermarking constraint, we check its impact to the satisfiability of the problem. If we detect that a to-be-added clause has the tendency to change a satisfiable formula to unsatisfiable, we may decide not to add it or to modify it first. In the rest of this section, we explain this in detail via three optimization-intensive watermarking techniques for the SAT problem. Similar idea can be applied for the protection of optimization problems to preserve the quality of the solution, where we only embed constraints that are very unlikely to change the optimality of the watermarked solutions.

3.2 SAT in EDA and SAT Solvers

Automatic test pattern generation (ATPG) is perhaps the most well-known application of SAT problem in EDA [100, 108, 153]. A combinational circuit can be represented by a function in the CNF format called the characteristic function [100]. A circuit is functionally consistent if and only if its characteristic function, a SAT formula, is satisfied. The characteristic functions of the simple gates are shown in Table 3.6. For a combinational circuits, one can set up a

Gate Type	Gate Function	Characteristic Function
NOT	$z =$ NOT(x)	$(x+z)(x'+z')$
AND	$z =$ AND$(x_1, \cdots, x_n)$	$\prod_{i=1}^{n}(x_i+z') \cdot (\sum_{i=1}^{n} x_i' + z)$
OR	$z =$ OR$(x_1, \cdots, x_n)$	$\prod_{i=1}^{n}(x_i'+z) \cdot (\sum_{i=1}^{n} x_i + z')$
NOR	$z =$ NOR$(x_1, \cdots, x_n)$	$\prod_{i=1}^{n}(x_i'+z') \cdot (\sum_{i=1}^{n} x_i + z)$
BUFFER	$z =$ BUFFER(x)	$(x+z')(x'+z)$

Table 3.6. Characteristic functions for simple gates[100].

variable for each node and conjuncts their characteristic functions together to obtain a characteristic function that represents the circuits. For example, the circuit in Figure 3.11 can be characterized by

$$\begin{aligned} f(x_1, x_2, x_3, y_1, y_2, z) &= (x_1 + y_1)(x_1' + y_1') \\ &\quad (x_2 + y_2')(x_3 + y_2')(x_2' + x_3' + y_2) \\ &\quad (y_1' + z)(y_2' + z)(y_1 + y_2 + z') \end{aligned}$$

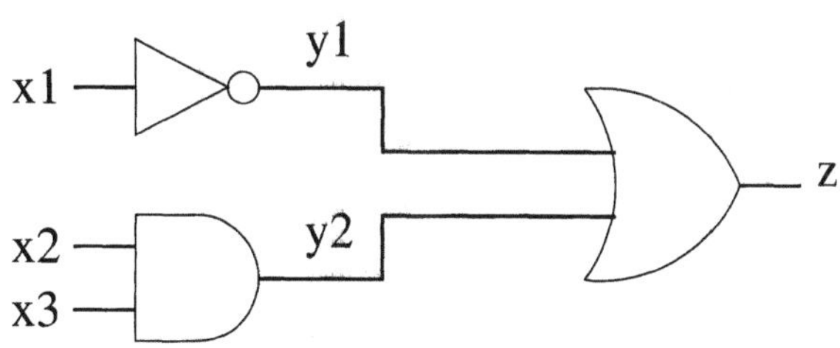

Figure 3.11. An example combinational circuit showing the characteristic function representation.

$f(x_1, x_2, x_3, y_1, y_2, z)$ is equal to one if and only if we have a valid assignment to all the variables: inputs x_1, x_2, x_3, output z, and the two intermediate output y_1, y_2. A stable state of the circuit must have its input/output satisfy this formula. To test whether we can have an output $z = 1$ with input $x_1 = 1$, all we need to do is to add two more one-literal clauses z and x_1 to f and solve the augmented SAT formula. (It is easy to see that in this case, when $x_1 = 1$, we have to assign both x_2 and x_3 to be 1 to make the output $z = 1$.).

Besides testing, researches have used SAT to solve many other problems in electronic design automation. To name a few, we mention FPGA routing in physical design [114, 149], logic synthesis [51], crosstalk noise analysis [33], and circuit delay computation [111].

Because of the importance of SAT in both theoretical and applied computer science, many heuristics have been developed to solve the problem[173, 107] and rigorous analysis has been conducted based on well-defined random models [58, 32]. The former gives us tools to solve the problem and the latter provides the theoretical background. Most of the current available SAT solvers fall into three categories:

Systematic search: The search process iterates through three steps: decision process that extends the current assignment by making a decision assignment to an unassigned variable; deduction process that extends the current assignment by following the logical consequences of the assignments made thus far; backtracking to undo the current assignment if it is conflicting, and trying another assignment. State-of-the-art solvers on this type: POSIT, NTAB, REL SAT and REL SAT-rand, Satz and Satz-rand. These solvers can handle up to 350 variable hard random formulas, while the hard 450 variable formulas are undoable[146].

Stochastic local search: The state-of-the-art stochastic solvers are GSAT and WalkSat. The basic idea behind these solvers is to pick a random initial assignment and then iteratively change the assignment of the variable that leads to the largest increase in the total number of satisfied clauses[148]. These solvers can solve hard formulas of 10,000 variables, but when it returns not satisfiable, it simply means a satisfying assignment was not found.

Translation to 0-1 integer programming: It is straightforward to translate SAT problems into 0-1 integer programming problems. However, currently the integer programming techniques cannot be made practical for satisfiability testing[148].

In recent years, many dedicated SAT algorithms have been developed targeting the large SAT instances from EDA domain [12, 107, 168]. They all fall into the systematic search category which has been proven effective for solving EDA applications, in particular for unsatisfiable instances. These algorithms are able to analyze the reasons of conflicts and conduct recursive learning during the search process.

3.3 Watermarking in the Optimization Fashion

The essence of the constraint-based watermarking method is to cut the solution space by adding extra constraints into the design process of the original IP. Then when we solve the watermarked problem (some overhead may be introduced as explained before), we only obtain solutions from the remaining solution space, i.e., those that satisfy both the additional constraints as well as the initial problem (c.f. Figure 3.2). The authorship is proved by showing the small probability for a random solution to satisfy all the extra constraints generated from the author's signature.

Obviously there is a trade-off between overhead and credibility, the two most important measures for a watermarking technique. Briefly, the tighter the extra constraints, the more difficult to solve the optimization problem, and hence the more degradation the quality of solution may suffer. However, they provide higher credibility in general as we have seen in the GC example.

For most optimization problems, we are guaranteed the existence of valid solutions despite of their quality. For example, any graph of n vertices is n-colorable in the graph vertex coloring problem, and there always exist tours in the traveling salesman problem. Therefore when we watermark optimization problems, our only concern is to keep the overhead as low as possible.

The decision problems, on the other hand, have only two different solutions: **YES** or **NO**. A formula is either *satisfiable* or *unsatisfiable* in the satisfiability problem and a graph does either contain or does not contain a subgraph isomorphic to another graph in the subgraph isomorphism problem. If the answer is YES, often a truth assignment or an isomorphic subgraph is required. When a decision problem has one unique answer, (e.g., an unsatisfiable SAT instance or a satisfiable instance with only one truth assignment), the solution space is so small that nothing can be hidden and therefore this technique fails. In general, for the constraint-based watermarking technique to be effective, we take the following "Watermarking Assumption" (Figure 3.12).

Watermarking Assumption: The decision problem to be watermarked must have an answer YES and have many different ways to achieve this answer.

Figure 3.12. Assumptions for decision problem watermarking.

This basic assumption corresponds to the "large solution space" requirement for the constraint-based watermarking on optimization problems. Since the watermarked IP has to maintain the correct functionality, i.e., the YES/NO answer in case of the decision problem, the question arises immediately when we add a watermark as extra constraints to the original problems: *Will the YES/NO answer stay unchanged as we watermark the decision problem?*

It is not difficult to construct counter-examples where we may turn a satisfiable formula to unsatisfiable by adding clauses, and find a graph contains a subgraph isomorphic to any other graph by introducing new edges and/or vertices. Under the "watermarking assumption", adding constraints may cut the solution space. It may happen that after the signature is completely embedded, we will get NO as the answer to the watermarked problem. To avoid this scenario, we propose the optimization version of the constraint-based watermarking, where only part of the signature is embedded.

The idea of optimization-intensive watermarking comes from an observation when we look at the essence of the methodology of constraint-based watermarking. The purpose of a watermark is to provide evidence of authorship and this is achieved by showing the small probability of coincidence that a random solution to the initial problem meet all the signature-based constraints. However, a 100% of authorship is never possible even if a perfect matching is found in the IP with the owner's signature because of the non-zero coincidence. In fact, we prove by reasoning that the probability of coincidence is so small that it is unlikely to happen. So there is no reason to embed the entire signature as long as we can provide a convincing proof of authorship.

We create a set of constraints from the to-be-embedded watermark. Each constraint makes some solutions invalid, and the constraints do not have the same effect in cutting the solution space. For example, the formula $x_1(x_2+x_3')(x_4+x_5'+x_6)$ can be easily satisfied, and it is still satisfiable after we add new clauses like $x_4(x_2+x_6')(x_1+x_2'+x_4')$, but it turns immediately to unsatisfiable if we add x_1'. For hard decision problems, there is no simple test that tells us which constraint will cut the solution space slightly and which one may completely change the answer to the problem. In the optimization constraint-based watermarking techniques we will present soon, we intend to add a subset of the constraints from the signature into the IP, based on statistical information while optimally keeping the YES/NO answer to the original decision problem.

3.4 Optimization-Intensive Watermarking Techniques for SAT Problem

In this part, we present three watermarking techniques on the satisfiability problem to explain the methodology of optimization-intensive constraint-based watermarking for decision problems.

Basic Notations:

$\{x_i : i = 1, 2, \ldots, n\}$ is a set of *boolean variables*, and we denote a variable x_i's *complement* by x_i'.

A *literal* is either a variable or its complement.

A *clause* is a disjunction (logic-OR, denoted by +) of one or more literals. We say a clause is true if and only if at least one of its literals is assigned value 1.

A *formula* is a conjunction (logic-AND, denoted by · or omitted when there is no ambiguity) of one or more clauses. A formula is satisfiable if there is a truth assignment to the variables, such that all the clauses are true.

Finally, for the simplicity of our analysis, we allow *redundancy* in the formula, i.e., one variable may appear multiple times in the same clause and a clause can occur in the same formula more than once. We call them *a generalized clause* and *a generalized formula*. Therefore, $x_1(x_1 + x_2 + x_2')x_1$ is a legal formula (which is functional equivalent to a single variable formula x_1 over two variables) under our definition. For example, the formula $\mathcal{F} = x_1(x_2 + x_3')(x_4 + x_5' + x_6)$ over variables $\{x_1, x_2, x_3, x_4, x_5, x_6\}$ is satisfiable and one truth assignment can be $\{x_1 = 1, x_2 = 1, x_3 =?, x_4 =?, x_5 = 0, x_6 =?\}$, where ? stands for *don't care* which means that the value of this variable does not affect the satisfiability of the given formula.

3.4.1 Adding Clauses

Given a set of n boolean variables, we may have 2^n truth assignments, this is the potential solution space of any satisfiability problem over this set of variables. A satisfiable formula has non-empty solution space while a unsatisfiable formula's solution space is empty. Any clause in a formula is a constraint that will prune the solution space. For instance, clause $x_1 + x_2$ will eliminate all truth assignments that assign both x_1 and x_2 to be 0 and hence cut one quarter of the solution space.

In the constraint-based watermarking process, a signature is embedded into the original problem as additional constraints to limit the choice of solutions. The natural constraint in the SAT problem is the clause and therefore the most straightforward way to embed signatures is to add new clauses. The extra clauses will be generated from the signature and any watermarked truth assignment will satisfy both the initial clauses as well as these signature-based ones. It is the fact that the additional clauses are met which is used to prove the existence of the signature. There are various ways to interpret a signature into extra clauses, Figure 3.13 shows one of them:

One important part is the calculation of the objective function, which we will discuss in details after we present the other two techniques. The introduction of objective function and a selective embedding distinguish the new optimization-intensive watermarking technique from the traditional "blind encoding" which embeds all of the signature. The objective function takes clauses as input and return a non-negative value, which measures the likelihood that adding these clauses will not change the satisfiability of the formula. As we explained before, it is impossible to construct such an objective function that tells exactly which

Input: a formula $\mathcal{F}$ over boolean variables $\{x_1, \ldots, x_n\}$,
and a message $\mathcal{M}$.
Output: a new formula $\mathcal{F}'$ derived from $\mathcal{F}$ with $mathcalM$ embedded.

Algorithm:

```
copy F to F';
convert M to a binary string S;
while ( S is not empty )
{ get the length l of the next clause C from the first k bit s of current S;
  construct the clause C from the next l⌊1 + log2 n⌋ bits of S;
  evaluate the objective function Obj at C;
  if ( Obj(C) ≥ preset_threshold ) add C to F';
  advance S;
}
report formula F';
```

Figure 3.13. Pseudo code for SAT watermarking: adding clauses.

clauses may change the answer to the formula. We have to test the satisfiability based on the statistic information of the formula.

3.4.2 Deleting Literals

In general, the longer the clause is, the easier it will be satisfied. (A clause with k literals is false if and only if all k literals are assigned 0). Based on this observation, we propose the second watermarking technique:

Input: a formula $\mathcal{F}$ over boolean variables $\{x_1, \ldots, x_n\}$,
and a message $\mathcal{M}$.
Output: a new formula $\mathcal{F}'$ derived from $\mathcal{F}$ with $mathcalM$ embedded.

Algorithm:

```
1. convert M to a binary string S;
2. declare a new formula F' with no clauses;
3. while ( S is not empty )
4. { get the current clause C from F;
5.   get the first ⌊log2 l⌋ bits from current string S, where l is the length of clause C;
6.   delete the k^th literal from C and get C', where k = ⌊log2 l⌋ in binary;
7.   evaluate the objective function Obj at clause C';
8.   if (Obj(C') ≥ preset_threshold)
9.     append C' to F';
10.    else append C to F';
11.    advance both F and S;
12. }
13. report formula F';
```

Figure 3.14. Pseudo code for SAT watermarking: deleting literals.

For example, let

$$\begin{aligned}\mathcal{F} &= (x_1)(x_3+x_5)(x_1'+x_5)(x_3+x_6+x_7'+x_{10})(x_2+x_6'+x_7)\\ &\quad (x_1'+x_2'+x_3'+x_4')(x_1'+x_2+x_5'+x_{10})(x_1'+x_3'+x_7+x_8+x_9')\\ &\quad (x_1+x_5'+x_7)(x_1'+x_2'+x_3+x_4+x_6+x_7'+x_9+x_{10}')\end{aligned}$$

And we want to embed the message "*June 1999*", which is 011011111001111 in binary where the first four digits represent the month (06) and rest for the year (1999). A non-optimization version of the above technique, as shown in Figure 3.14 without lines 7, 8, and 10, will skip the evaluation of the objective function and simply append every new clause $\mathcal{C}$ to $\mathcal{F}'$. In this example, literals $x_3, x_5, x_7', x_6', x_4', x_{10}, x_1', x_5'$ and x_{10}' will be deleted respectively from $\mathcal{F}$ starting with the second clause:

$$\begin{aligned}\mathcal{F}' &= (x_1)(x_5)(x_1')(x_3+x_6+x_{10})(x_2+x_7)\\ &\quad (x_1'+x_2'+x_3')(x_1'+x_2+x_5')(x_3'+x_7+x_8+x_9')\\ &\quad (x_1+x_7)(x_1'+x_2'+x_3+x_4+x_6+x_7'+x_9)\end{aligned}$$

Formula $\mathcal{F}'$ has exactly the same number of clauses as $\mathcal{F}$ but with one literal less in each clause (except for single-literal clauses). It is clear that the solution space of $\mathcal{F}'$ is a proper subset of that for $\mathcal{F}$, so any truth assignment that satisfies $\mathcal{F}'$ also satisfies $\mathcal{F}$. However, we see that in this case $\mathcal{F}'$ is unsatisfiable because of the single-literal clauses x_1 and x_1'. Therefore, the traditional method fails.

As illustrated in Figure 3.14, in the proposed optimization-intensive watermarking process, the strength of each additional constraint is estimated before it is embedded. In this case, for example, it may detect that after deleting literal x_5 from the third clause $(x_1' + x_5)$, the remaining (single-literal) clause x_1' can hardly be satisfied, i.e., $Obj(x_1') \ll$ preset_threshold, and thus the original clause $(x_1' + x_5)$ is kept. For the same reason, the deletion of x_4' from the sixth clause is ignored and we get an optimization-intensive watermarked SAT instance, which is still satisfiable:

$$\begin{aligned}\mathcal{F}'' &= (x_1)(x_5)(x_1'+x_5)(x_3+x_6+x_{10})(x_2+x_7)\\ &\quad (x_1'+x_2'+x_3'+x_4')(x_1'+x_2+x_5')(x_3'+x_7+x_8+x_9')\\ &\quad (x_1+x_7)(x_1'+x_2'+x_3+x_4+x_6+x_7'+x_9)\end{aligned}$$

3.4.3 Push-out and Pull-back

The constraint-based watermarking techniques add signature-related constraints to the original problem, cut its solution space and thus increase the chance of getting a watermarked solution. When these additional constraints are too strong to keep the quality of the solution, we introduce the optimization-intensive technique to embed the constraints in a selective way, which excludes the addition of "bad" constraints. The previous "Adding Clauses" and "Deleting

Literals" techniques work on the original solution space and try to make "good" decisions on embedding a constraint or not. Hence there are natural limitations imposed by the SAT instance itself. In "Adding Clauses", no more constraints can be added when there is only one truth assignment left. In "Deleting Literals", removing all literals from a clause eliminates one original constraint and may result in wrong solutions. The third technique we propose here breaks this barrier by a two-phase push-out and pull-back procedure.

In the push-out phase, the solution space is enlarged such that there will be more room to hide the signature. For SAT problem, this can be done by either introducing new variables (and clauses) or deleting clauses. As we discussed earlier, deletion of clauses cannot preserve the validity of the solution and therefore we focus on introducing new variables. When we treat the SAT instance as a formula over the initial set of variables and a new variable x, the solution space is doubled because x is not involved in the formula and will serve as a "don't care" variable. It is in this larger solution space that we apply various (optimization-intensive) watermarking schemes to embed the signature and create a (optimization-intensive) watermarked SAT instance. Once we solve such instance and get a solution over the extended set of variables, we can restrict the truth assignment to the initial variables and the extended solution is pulled back. This is illustrated in Figure 3.15, where the shaded area in (c) and (d) is the solution space for the watermarked formula.

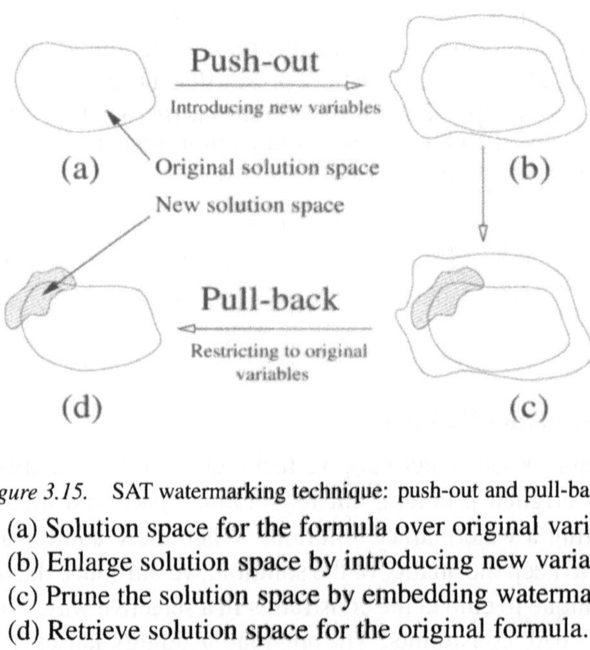

Figure 3.15. SAT watermarking technique: push-out and pull-back.
(a) Solution space for the formula over original variables.
(b) Enlarge solution space by introducing new variables.
(c) Prune the solution space by embedding watermark.
(d) Retrieve solution space for the original formula.

This technique can be combined with the previous ones and yields more powerful watermarking method. For example, with the freedom of adding new variables, we can change the "adding clauses" technique in the following way: whenever we detect a dangerous clause, i.e., one that may make the entire formula unsatisfiable, we introduce a new variable to the clause. In this way, we have better chance to maintain the satisfiability of the watermarked formula, and we can build new clauses over the increased variable set.

3.5 Analysis of the Optimization-Intensive Watermarking Techniques

We first show the correctness of the proposed watermarking techniques, then discuss the objective function we mentioned in the previous section. We analyze the limitation of these techniques on one widely-used SAT model and conclude with a discussion on how to detect a watermark from a given solution to a formula.

3.5.1 The Correctness of the Watermarking Techniques

Let $\Psi = \{\mathcal{F} : \mathcal{F} \text{ is a formula over a set of boolean variables } \{x_1, x_2, \cdots,\}\}$, we first define a partial order "$\preceq$" on Ψ and say formula $\mathcal{F}_1$ is *more constrained* than $\mathcal{F}_2$ if the partial order $\mathcal{F}_1 \preceq \mathcal{F}_2$ holds:

Definition 5.1: For two clauses $C_1 = l_{11} + l_{12} + \cdots + l_{1n_1}, C_2 = l_{21} + l_{22} + \cdots + l_{2n_2}$, denote $C_1 \preceq C_2$ iff $\{l_{11}, l_{12}, \cdots, l_{1n_1}\} \subseteq \{l_{21}, l_{22}, \cdots, l_{2n_2}\}$, i.e., $\forall\, i \in \{1, 2, \cdots, n_1\}$, $\exists\, j \in \{1, 2, \cdots, n_2\}$, such that $l_{1i} = l_{2j}$. And for two formulas $\mathcal{F}_1 = C_{11} \cdot C_{12} \cdots C_{1n_1}, \mathcal{F}_2 = C_{21} \cdot C_{22} \cdots C_{2n_2}$, we define $\mathcal{F}_1 \preceq \mathcal{F}_2$ iff $\forall\, j \in \{1, 2, \cdots, n_2\}, \exists i \in \{1, 2, \cdots, n_1\}$, such that $C_{1i} \preceq C_{2j}$.

It is clear that the above defines a partial order. Given two formulas $\mathcal{F}$ and $\mathcal{F}'$ with $\mathcal{F} \preceq \mathcal{F}'$, then for every clause C' (constraints to the SAT instance) in $\mathcal{F}'$, there exists a clause C in $\mathcal{F}$ such that C' will be satisfied whenever C is. I.e., $\mathcal{F}$ has all the constraints that $\mathcal{F}'$ has.

When a signature is added as extra clauses, the watermarked formula will become more constrained than the original one and therefore any watermarked solution will remain valid. For "Deleting literals", when a literal is eliminated from a clause, that clause becomes more constrained and so will be the watermarked formula. In sum, we have the following observations:

Proposition 5.2: If $\mathcal{F} \preceq \mathcal{F}'$ and $\mathcal{F}$ is satisfiable, then $\mathcal{F}'$ is also satisfiable. Moreover, any truth assignment to $\mathcal{F}$ satisfies $\mathcal{F}'$.

Proposition 5.3: Let $\mathcal{F}'$ be a (optimization-intensive) watermarked formula from an original formula $\mathcal{F}$, then $\mathcal{F}' \preceq \mathcal{F}$. Hence any watermarked truth assignment to $\mathcal{F}'$ meets the requirement of $\mathcal{F}$.

3.5.2 The Objective Function

An objective function $Obj : \Psi \rightarrow R^+ \cup \{0\}$ measures the likelihood that a formula can be satisfied. Ideally, any objective function Obj should assign unsatisfiable formulas a value of 0, easy SAT instances larger values, and be non-decreasing over the partial order $(\Psi, \preceq)$. I.e., $Obj(\mathcal{F}) \leq Obj(\mathcal{F}')$ for any formulas $\mathcal{F} \preceq \mathcal{F}'$. For example, it can be defined as:

- $Obj(\mathcal{F}) = min_{1 \leq i \leq n}\{Obj(C_i)\}$ for any formula $\mathcal{F} = C_1 \cdot C_2 \cdots C_n$.
- $Obj(C) = \Sigma_{i=1}^{n} Obj(l_i)$ for any clause $C = l_1 + l_2 + \cdots + l_n$.
- extend the notation by denoting $Obj(l)$ the likelihood that literal l is assigned true.

The only part left to be specified is how to determine the values of $Obj(l)$ and $Obj(l')$ for a literal l and its complementary l'. Intuitively, the more often a literal l appears in the formula and the less its complementary occurs, l will have better chance to receive true. Let n_l be the number of occurrence of l and we can finish the definition:

Zero order objective function

$$Obj_0(l) = \begin{cases} \frac{f_0(l)}{f_0(l')} & n_{l'} \neq 0 \\ \infty & n_{l'} = 0, n_l \neq 0 \\ \text{not defined} & n_{l'} = 0, n_l = 0 \end{cases} \tag{3.6}$$

where

$$f_0(l) = n_l \tag{3.7}$$

Basically, Equation (3.6) uses the ratio of the occurrences as the measurement for the assigning variables true/false. If l's complementary form l' never appears in the formula, to find a truth assignment, it does not hurt us at all to make l true. And if the formula does not contain a particular variable, there is no need to define the objective function on this variable.

First order objective function

From the zero order objective function, we see that every occurrence of l will increase $Obj(l)$ and decrease $Obj(l')$. However, the contribution of each occurrence is related to the length of the clause and this is not considered in the zero order objective function. The literal in any single-literal clause has to be assigned true and the value of any particular literal is not that crucial for a clause with many literals. For a literal l, let n_l be the number of clauses that contains l and s_i be the length of the ith such clause. Then we define the first order objective function on l as:

$$Obj_1(l) = \begin{cases} \frac{f_1(l)}{f_1(l')} & n_{l'} \neq 0 \text{ and } f_1(l') \neq \infty \\ \infty & n_{l'} = 0, n_l \neq 0 \\ \text{not defined} & n_{l'} = n_l = 0 \text{ or } f_1(l) = f_1(l') = \infty \end{cases} \tag{3.8}$$

where

$$f_1(l) = \sum_{i=1}^{n_l} \frac{2^{s_i-2}}{2^{s_i-1}-1} \tag{3.9}$$

There are $2^k - 1$ distinct truth assignments for a clause of length k, out of which 2^{k-1} will have a particular literal assigned true. Equation (3.9) is a simple modification of this fact which enforces f_1 to evaluate to ∞ at literals from the single-literal clauses. From this definition, it is easy to verify that:

Proposition 5.2: The first order objective function satisfies:

(i) $Obj(l) = \infty$ iff the formula does not contain l' or has l, but not l', as a single-literal clause.

(ii) $Obj(l)$ is increasing with respect to n_l and decreasing with respect to $n_{l'}$.

(iii) $Obj(l)$ is decreasing with respect to s_i.

(i) implies that if the formula does not have l' or has l as a single-literal clause, then setting l true only helps us finding a solution. When the formula has both l and l' as single-literal clauses, obviously it is unsatisfiable; (ii) suggests that the more l occurs, the more likely it will be assigned true; and (iii) says the longer is the clause, the less it contributes to the objective function since a long clause is easy to satisfy.

Second order objective function

Although the function Obj_1 is better than Obj_0 in describing the likelihood of a literal being assigned true, by no means it is most accurate. Considering two clauses: $C_1 = l_1+l_2+l_3$ and $C_2 = l_1+l_4+l_5$, C_1 and C_2 will contribute the same amount, $\frac{2}{3}$, to $f_1(l_1)$ by Equation (3.9). However, this becomes inaccurate if we know, from the rest part of the formula, that most likely l_2 or l_3 will be true while both l_4 and l_5 are false. Where $f_1(l_1)$ should receive a large boost from clause C_2 and little from C_1. This suggests us that we should also study the correlation between literals. By modifying $f_1(l)$, we can define the second order objective function in a similar way with:

$$f_2(l) = \sum_{i=1}^{n_l} \frac{1}{2^{s_i-1} - 1 + \Sigma_{j=1,l_j\neq l}^{s_i} Obj_1(l_j)} \tag{3.10}$$

The purpose of introducing objective functions is to provide criteria that can be used to determine whether an additional constraint should be embedded or not during the optimization-intensive watermarking process. An objective function estimates the difficulty of determining the satisfiability of a formula. Obj_0 considers only the occurrence of l, l' and uses the ratio as a measure. Obj_1 takes into account the length of each clause that a literal and its complementary appears. In the second order objective function, not only l and l, but also their *neighbors* (the literals in the same clause) are considered. Therefore it provides more accurate estimation. Of course, better objective functions can be defined when we use more information from the SAT problem. Unfortunately, since the objective function will be called frequently, the computation cost of such function should be as low as possible. Usually, the accuracy of the objective function is at the expense of its complexity. For example, both Obj_0 and Obj_1 can be computed when the SAT instance is read in with the help additional storage. However, one more parse of the SAT instance is required to initialize Obj_2. A perfect objective function should be able to tell exactly the satisfiability of an instance and it cannot be computed in polynomial time unless $P = NP$.

For a given satisfiable formula, the optimization watermarking techniques do not guarantee the watermarked formula still satisfiable, but maximize this probability. Before we discuss the limitation of the proposed techniques, we mention a couple of properties of the defined objective functions:

$$\begin{aligned}
Obj(\mathcal{F}) = 0 &\Leftrightarrow \mathcal{F} \text{ is } \textit{unsatisfiable} \\
Obj(\mathcal{F}) = \infty &\Leftrightarrow \mathcal{F} \text{ is trivially } \textit{satisfiable} \\
Obj(l) = \infty &\Rightarrow \mathcal{F} \text{ is satisfiable if } \mathcal{F} \text{ is satisfiable by assigning } l \text{ to be true}
\end{aligned}$$

3.5.3 Limitations of the Optimization-Intensive Watermarking Techniques on Random SAT

The constant-probability SAT model

We adopt the model $J(n, r, p)$ for generating random SAT instances. A formula of this type consists of n clauses of r variables. A variable v is in the kth clause as an uncomplementary literal with probability p, as a complementary literal with probability p, and the kth clause does not contain variable v with probability $1 - 2p$.

Franco and Ho[58] proved that, for this model, almost all SAT instances can be solved in polynomial time if any of the following conditions holds:

$$p < \frac{0.4 \ln n}{r} \tag{3.11}$$

$$p > \frac{\ln n}{r} \tag{3.12}$$

$$p = c\frac{\ln n}{r} \text{ and } \lim_{n,r\to\infty} \frac{n^{1-c}}{r^{1-\varepsilon}} < \infty, (0.4 < c < 1, \varepsilon > 0) \quad (3.13)$$

It is also shown that almost all randomly generated SAT have no solution if:

$$p = c\frac{\ln n}{r}, \text{ and } \lim_{n,r\to\infty} \frac{n^{1-c}}{r} = \infty\, (\, 0.4 < c < 1\,) \quad (3.14)$$

Figure 3.16 [58] shows the relationships between the parameters of model $J(n, r, p)$ that result in random instances that are always solvable in polynomial time. Curve I represents $e^{\frac{pr}{0.4}}$ and the region to the left of it (Equation (3.11)) are instances that are always unsatisfiable due to the large amount of clauses. Curve II s e^{pr} and the region to its right (Equation (3.12) corresponds to instances that are almost always satisfiable. According to Equation (3.14), the instances above curve III are almost unsatisfiable. The shaded area is a mixture of satisfiable and unsatisfiable problems.

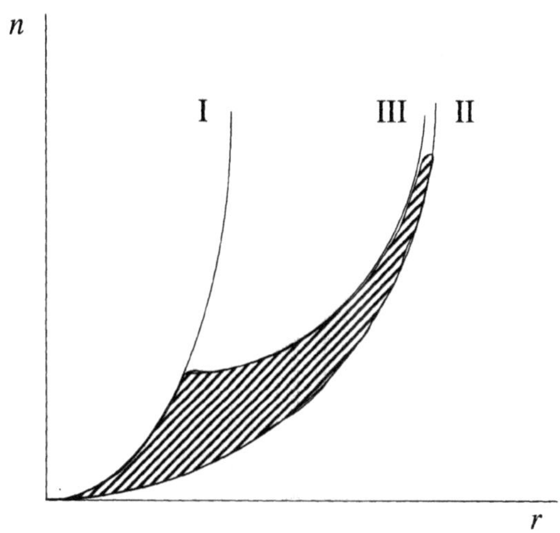

Figure 3.16. The satisfiability of model $J(n, r, p)$ (redrawn from [58]).

Limitations on the optimization techniques

Under the “watermarking assumption", a to-be-watermarked SAT instance belongs to the region right to curve II as shown in Figure 3.17(a), where the solution space is large. After we embed the signature, the SAT instance and/or the curves may change. We do not want the new instance to fall in the area left of curve I or above curve III, where the probability that the new instance is unsatisfiable is almost 1. Even for a satisfiable watermarked instance in the shaded region, it usually becomes hard to find a truth assignment. We now graphically analyze the impact of the proposed watermarking techniques.

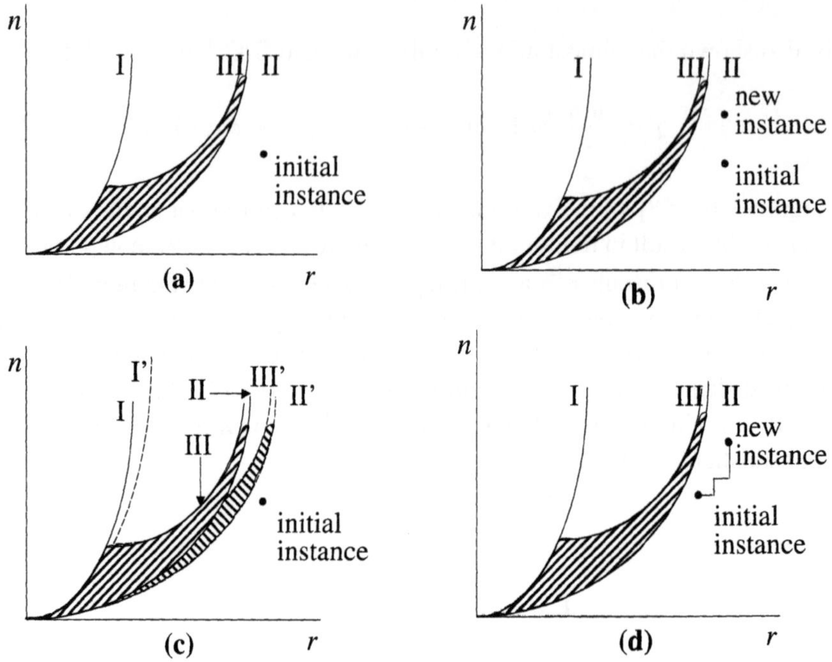

Figure 3.17. A SAT instance and its watermarked versions. (a) The initial SAT instance; (b) New instance afteradding clauses; (c) New instance (same spot as initial) and new curves after deleting literals; (d) New instance after push-out and pull-back.

Adding clauses: Assuming the message is random, and the length of a new clause is chosen in accord with the initial instance, then the watermarked instance is still a random SAT problem of the same type, except that the number of clauses has increased. This is shown in Figure 3.17(b), where curves I, II, and III remain the same, the new instance is right above the initial one, which indicates an increment of n with the same r and p. It is clear that if we keep on adding new clauses, the watermarked instance will cross curve II, making the instance hard to solve and eventually becomes unsatisfiable.

Deleting literals: If we delete literals based on a random message, our optimization strategy will keep us from deleting single-literal clauses and eliminating any variable completely from the formula. Therefore the new instance will be a formula on the same set of variables with the same number of clauses. In the r-n chart (Figure 3.17(c)), the new instance shares the same position as the initial one. However, all the curves have moved towards right because of the decrement of p due to the deletion of literals. When there are only few literals left, p will become extremely small and all the curves will cross the SAT instance and make it unsatisfiable.

Push-out and pull-back: In this technique, new variables only appear in the clauses corresponding to the signature, so it is not appropriate to use the same model. However, the idea can be illustrated by Figure 3.17(d), the initial instance is moving along r-axis as we add new variables, then moving up as we append new clauses. New variables are introduced whenever the new instance moves close to curve II and the addition of a new variable keeps the watermarked formula in the region under the "watermarking assumption". Technically, there is no limitation on this technique if any number of new variables can be added.

3.5.4 Copy Detection

Detecting copies is one of the fundamental problems for distributing IPs among users. An embedded watermark is useful only if the IP provider can detect it and prove his/her authorship to the third party, which is the sole goal of copy detection. Our key idea used to protect the SAT solution is to prune the solution space based on the signature and then get the solution from this small space. The strength of the authorship depends on the size of the solution space for the watermarked problem relatively to the original one. Here we outline the approaches to retrieve watermarks embedded by the "adding clauses", similar results hold for the other techniques.

In the "adding clauses" method, the solution is forced to satisfy extra clauses according to the signature. Suppose the signature is translated to k clauses of length $w_1, w_2, \ldots, w_k$ respectively. Let

$$P_k = \Pi_{i=1}^{k} 1 - (\frac{1}{2})^{w_i} \tag{3.15}$$

Then we have:

Proposition 5.3: A random assignment makes all k clauses true with a probability P_k, and the probability that it satisfies at least $m < k$ clauses is:

$$P_m = P_k\{1 + \sum_{i_1=1}^{k} \frac{2^{-w_i}}{1-2^{-w_i}} + \sum_{i_1 \neq 12, i_1, i_2 = 1}^{k} \frac{2^{-(w_{i_1}+w_{i_2})}}{(1-2^{-w_{i_1}})(1-2^{-w_{i_1}})} + \ldots + \sum_{i_j \text{ distinct}, i_1,\ldots,i_{k-m}=1}^{k} \frac{2^{-\Sigma_{j=1}^{k-m} w_{i_j}}}{\Pi_{j=1}^{k-m} 1 - 2^{-w_{i_j}}}\}.$$

Corollary 5.4: For 3-SAT, where all the clauses are of length 3,

$$P_m = (\frac{7}{8})^k \sum_{i=0}^{k-m} \binom{k}{m} (\frac{1}{7})^i \tag{3.16}$$

It is easy to see for the expression of P_m that this probability can be arbitrarily small when both k and m are large enough. Thus, this method provides high credibility for signatures of large instances. In practice, for a given SAT instance, from the limitation of the technique we can determine the maximal constraints we may introduce. Then according to the level of credibility we want to achieve, we can calculate the minimal constraints we have to add to the original problem and then fine tune the objective function.

Any clause, which is independent of the original formula $\mathcal{F}$, of length i will be satisfied by a truth assignment to $\mathcal{F}$ with probability: $1 - (\frac{1}{2})^i$. Hence, the entire watermark can be satisfied with probability:

$$\Pi_{i=1}^{k} 1 - (\frac{1}{2})^{w_i}$$

On the other hand, the solution provided by solving the watermarked formula *F'* will satisfy all the extra clauses with a much higher probability, which depends on the implementation of the technique, and in extreme, if a truth assignment is found without using the optimization, the entire watermark will be guaranteed satisfied.

Alternatively, we can prove the authorship by showing, among the k watermarked clauses, how many are satisfied by the truth assignment. Again, the watermarked solution is expected to satisfy much more.

3.6 Experimental Results

We have implemented our proposed optimization-intensive watermarking techniques and apply them to a set of instances from DIMACS SAT benchmarks [174].

File	Variables	Clauses	Ratio	Clause Length
ii8a1.cnf	66	186	2.4	$2 \sim 3$
ii8a2.cnf	180	800	2.6	$2 \sim 6$
ii8a3.cnf	264	1552	2.6	$2 \sim 6$
ii8a4.cnf	396	2798	2.7	$2 \sim 6$
ii8b1.cnf	336	2068	-	$2 \sim 6$
ii8b2.cnf	576	4088	2.3	$2 \sim 6$
ii8b3.cnf	816	6108	-	$2 \sim 6$
ii8b4.cnf	1068	8214	2.2	$2 \sim 6$
ii8c1.cnf	510	3065	2.4	$2 \sim 10$
ii8c2.cnf	950	6689	2.3	$2 \sim 10$

Table 3.7. Characteristics of benchmarks. "Ratio" is measured by literals/clauses and "Clause Length" is the range for the length of clauses.

The ii8*.cnf instances are generated from the problem of inferring the logic in an 8-input, 1-output "blackbox". We watermark each of these instances using regular techniques without optimization, then apply the optimization-intensive techniques to embed the same message. The results show that in most instances, much longer messages can be embedded by the new techniques before changing the problem to unsatisfiable. Both the initial and watermarked instances are solved by *WalkSAT*, a solver implemented by Kautz and Selman[173]. All instances are solved instantaneously, the run-time overhead is negligible.

Among the techniques we proposed, the "adding clauses" method has the best performance. We first generate a long random bit-stream as our message, then create clauses of variable length according to this message and append them to the original problem. Table 3.7 reports the maximal length of the bit-steam that we can take before turning the problem to unsatisfiable.

As one can see form Table 3.7, we achieve an average of 58.68% improvement. It is worth mentioning here that in the worst case, we successfully embedded 1400 bits, which corresponds to 63 clauses. Although the probability that a random assignment satisfies 63 additional clauses is not very small, ($< 2.3 \times 10^{-4}$), the chance that these clauses are created from a meaningful message is low.

Instance	w/o Optimization	Optimization-Intensive	Improvement Ratio
ii8a1.cnf	1900	1400	−26.32%
ii8a2.cnf	4900	3100	−36.73%
ii8a3.cnf	3900	6700	71.80%
ii8a4.cnf	3900	3900	0
ii8b1.cnf	3800	6800	78.95%
ii8b2.cnf	4900	8900	81.63%
ii8b3.cnf	4900	8900	81.63%
ii8b4.cnf	3900	11600	197.44%
ii8c1.cnf	3900	7500	92.31%
ii8c2.cnf	8900	13000	46.07%
Average Improvement			58.68%
Median Improvement			78.95%

Table 3.8. Improvement of the optimization-intensive technique over regular watermarking technique.

We also test the proposed methods on random 3-SAT instances, where the literal per clause ratio is fixed at 4.25. These instances are in the range of "hard-to-be-solved"[32]. Although all the problems are known to be satisfiable, it is not expected that many satisfying assignments exist. Therefore, the "watermarking assumption" does not hold. When we try to watermark these problems, very limited message can be embedded (less than 100 bits), and the optimization-intensive techniques do not help that much. (Imagine an instance very close to curve II in Figure 3.16).

4. Summary

In this Chapter, we propose a constraint-based watermarking technique for IP protection. Instead of solving the real problem and posting the answer directly, we build a watermarking engineer which takes the real problem and the owner's signature as input and gives a solution to the initial problem with the given signature embedded. Inside the watermarking engineer, we translate the signature into a set of additional constraints and add them into the original problem. Therefore, the solution will satisfy both the original and additional constraints. I.e., in this solution, there exist special structures that cannot be easily discovered without the owner's signature. Now the owner can claim his/her authorship by showing the small probability that such structures exist in a random solution without watermark.

Since the signature is embedded as extra constraints, there might be some degradation in the quality of the IP. The trade-off between *credibility* (measures for the strength of proof for authorship) and *overhead* (measures for the degradation of quality of the IP) has to be balanced. Besides, there are other requirements for a watermarking technique to be effective. We discuss these requirements and build a framework to evaluate different watermarking techniques. The analytical foundations we lay out here is valid for the analysis of all watermarking techniques, not only for those that we have discussed in this chapter.

We have also proposed the first set of optimization-intensive watermarking techniques for decision problems. The basic concept of these techniques is to select a subset of the signature and embed it as the watermark. Theoretically, we have showed that this partial signature will provide convincing authorship and an average of 58.68% improvement is achieved in practice when we implement this idea to watermark a set of benchmark SAT instances.

Figure 3.18 summarizes the current state of constraint-based watermarking techniques. The goal is to protect IPs that require to maintain the correct functionality and we achieve it by adding constraints during the design and implementation of the IP. The addition of signature-based constraints will not alter the initial constraints and therefore will keep the IP's functionality. However, the extra constraints enforce (watermarked) solutions to have rather unique structures which are used as proof of the authorship. Although this idea originally targets optimization problems, we propose the optimization-intensive technique to extend it for the protection of decision problems. Further improvements, such as fair watermarking, hierarchical watermarking, and local watermarking, are introduced as well to cover more specific concerns.

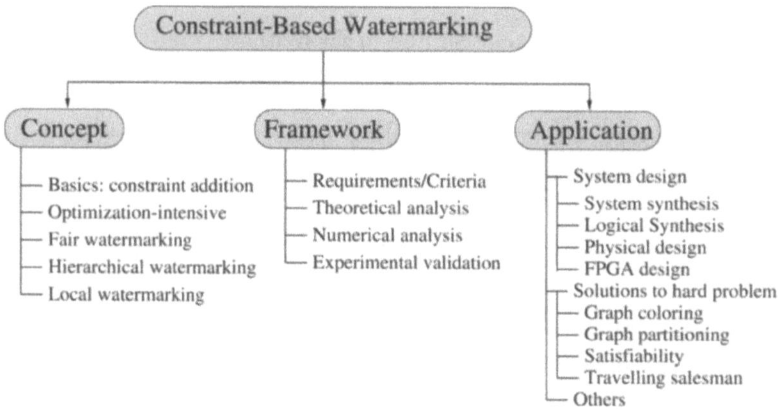

Figure 3.18. Outline of research on constraint-based watermarking.

We layout a set of requirements for the watermark to be effective, namely: correct functionality, low overhead, high credibility, resilience, transparency, part protection, and fairness. One can also use these as the criteria to compare difference watermarking approaches quantitatively. Then based on the model of coloring random graphs, we conduct both theoretical and numerical analysis. We show that low overhead and high credibility can be achieved at the same time, which makes a solid mathematical background for the constraint-based watermarking paradigm. Furthermore, we built testbed to validate our approach through experiments and simulations. On one hand, we apply the watermarking techniques on real life problems; on the other hand, we set up experimental platform (e.g., the SAT model with known solutions) to validate the new concepts like fairness in the context of IP protection.

Finally, we have seen a number of applications based on this idea. One of the most successful is the protection of system design process. Due to the natural hierarchical structure of design process, designers can embed watermark during each design stage. The watermarks in an earlier stage will be propagated to later stages and eventually embedded in the final IP. Examples can be found in FPGA design[98, 99], physical design[27, 81], logic synthesis[89], behavioral synthesis [75], DSP design[74], and so on. Another domain of applications is the protection of solutions to hard problems: graph coloring[132], graph partitioning[164], Boolean satisfiability[133, 135], and more recently, shortest path in maps[85].

Notes

1 The additional constraints are similar to those in the original graph, and we introduce them in a way such that original graph's characteristics will not be changes. For example, if the original graph is planar, any encoding scheme that adds constraint making the resulting graph non-planar will be bad.

2 Although the ASCII code of the signature file can be used as well, the digital encryption and pseudorandom bitstream enhance the security and credibility of this entire IP protection process. We will further discuss this in the next section.

3 In most of the graph coloring algorithms, two nodes that have many common neighbors will be assigned the same color. Let's call such two nodes A and B, now if A has a neighbor C that is connected to B, suppose A and B have already received the same color, then A and C will have different colors no matter we add an edge between them or not.

4 A clique is a subset of nodes such that they are all connected to each other. Clearly we need the same number of colors as the nodes in a clique to mark this clique.

5 We choose expression (3.3) instead of (3.2) to simplify the asymptotic analysis, all the results hold if we replace (3.3) by (3.2).

6 In general, the graph is not random unless k is a multiple of n. The randomness can be maintained by modifying this technique in the following way: in Figure 3.3, select the first vertex of each pair according to the message M instead of the given order for the vertices. E.g., the first node will be v_l where the binary expression of l: $l_2 = m_0 m_1 \cdots m_{\lfloor \log_2 n \rfloor}$. In practice, we restrict k to be multiples of n to keep the randomness.

7 For a given MIS of size k, selecting these k vertices in different orders delivers different messages. However, it is unlikely to get the same MIS from different messages (after encryption).

8 Alternatively, we can map long messages to a fixed length message by hash functions. Since hash function is many-to-one, this brings ambiguity which depends on the hash function itself. Such analysis is out of the scope of this paper.

9 For the last node, we can add edges randomly or repeat the message to make sure it has $(n + k - 1)p$ neighbors.

10 The encryption of the signature file and the development of the standard encoding scheme should be separated. An encoding scheme that depends on the signature file is suspicious and not convincing. Because for a given solution, one can deliberate a watermarking procedure that makes the solution corresponds to any signature.

Chapter 4

FINGERPRINTING FOR IP USER'S RIGHT PROTECTION

The goal of intellectual property protection is to ensure the rights of both the IP providers and the IP users. The watermarking-based approaches do not facilitate tracing of illegally resold IPs and therefore cannot provide protection for buyers. In this chapter, we present a generic symmetric fingerprinting technique which can be applied to an arbitrary optimization/synthesis problem and, therefore, to hardware and software IPs. Fingerprinting techniques require to issue different IP users distinct copies of the same IP, we also propose a zero run-time overhead fingerprinting method that provides us controllable number of distinct fingerprinted copies.

1. Motivation and Challenges

Today's engineering teams are facing more severe challenges than ever: the shortage of engineering manpower, the soaring design complexity, the growing time-to-market pressure, and the fast rising fabrication cost just to name a few. According to a study of 320 engineering teams in North America by Collett International [170], by the year 2000, the new-design productivity must be doubled and reuse productivity must be improved by a factor of 12. At the same time, design cycle time must drop by 15 percent, team size grows by 36 percent, and reuse increase by 53 percent.

Multi-vendor IP integration is by far the most promising solution to these challenges. As an evidence, we have seen CAD tool capability, IP-based design and reuse methodologies getting a great deal of industrial and academic interest. IP protection techniques are an unavoidable prerequisite for development and adoption of reuse-based system integration business models. In such reuse-based IP business models, as well as the related IP protection model, there are two basic types of legal entities involved in an IP transaction: *provider* (*seller*, *owner*) and *buyer* (*user*). The goal of IP protection is to protect the rights of both the provider and the buyer.

The ownership can be protected by the *constraint-based watermarking* technique and its derivatives. All of these techniques are based on the idea of embedding IP provider's signature as additional design constraints to create a rather unique IP. From the IP providers' standpoint, this is not enough to discourage piracy and unauthorized redistribution: the *buyer's* legal ownership of a given piece of IP must symmetrically be protected as well.

The IP provider desires the ability to trace a dishonest buyer from unauthorized resold copies of the IP. It is crucial for IP provider to distribute IPs with the same functionality but different appearance to different users. Because the problem of tracing traitors will become insurmountable if all users get exact the same IP and one of them illegally redistributes. This problem can be solved by embedding the IP provider's signature into the design (for protecting ownership), and additionally embedding a unique signature to each realization of the design (for tracing traitors and protecting legal users).

On the IP buyers' side, they also demand the protection from being "framed" by other dishonest users working in collusion, or by a dishonest provider who sells extra copies of the IP and then attempts to blame the buyer. The buyer can provide the IP provider with his signature which is encrypted using the buyer's public key. He can easily check whether the purchased design indeed contains this signature. Since the buyer is the only entity who can interpret the signature (using his secret key), he is also protected in the sense that now the provider can not resell the IP without the buyer's permission.

Such symmetric protection of the provider's and buyer's rights is afforded by a *fingerprinting* methodology, whereby the IP provider *fingerprints* and delivers to each buyer a unique copy of functionally identical IP. Fingerprinting schemes have been widely and effectively used to trace individual object. However, their application domain has been restricted only to static artifacts, such as image and audio where distinct copies can be easily created. There have been a lot of reported fingerprinting protocols for digital data sets[18, 20, 122]. Almost all of them make use of the end users' insensitivity of minute errors (for example, flip the lease significant bit) in the copies they receive. Clearly this is not applicable to generating IPs that require the correct functionality.

The main challenge in IP fingerprinting is how to implement the same IP, functionality-wise, in many different ways to accommodate the potential IP user market. One straightforward approach is to acquire each IP user's signature and repeat the entire design process to embed such signature. Creating a large number of different high-quality solutions from scratch has a clear time and cost overhead that the IP provider most often cannot afford. Therefore, we require fingerprinting protocols that can provide a number of distinct versions of the same IP with reasonable amortized design effort.

The first IP fingerprinting technique in the literature is due to Lach et al. [97]. Their approach is based on solution partitioning. By partitioning an initial solution into a large number of parts and by providing for each part several different realizations, one can realize a fingerprinting scheme with relatively low performance impact for their application (a restricted FPGA mapping problem). However, the technique of [97] cannot be applied to design steps that do not have natural geometric structure and that are sensitive to the cost of the solution. More importantly, the technique has relatively low resilience against collusion attacks since it produces solutions with identical global structure (cf. the work of Boneh and Shaw [20]). Finally, the time overhead associated with creating fingerprinted solutions is relatively high.

2. Fingerprinting Objectives

2.1 A Symmetric Interactive IP Fingerprinting Technique

Fingerprints are the characteristic of an object that is completely unique and incontrovertible. They have been used for human identification for a long time because of their uniqueness. Recently, many fingerprint sensor chips and systems have been developed[78, 113]. Protocols have been developed for adding fingerprint-like marks into digital data to protect both the provider and the buyers [18, 20, 125]. Boneh and Shaw [20] propose the most efficient symmetric fingerprinting schemes in the sense that both the distributor and the user know the fingerprinted copy. Pfitzmann and Schunter[125] introduce asymmetric fingerprints, where only the user knows the fingerprinted data while the distributor can identify the user's information from the data. Biehl and Meyer[18] combine these two and give a construction more suitable for broadcast data.

Like the watermarking techniques for artifacts, such fingerprint-like marks are made by introducing minute errors to the original copy, with such errors being so insignificant that their effect is negligible. All of these techniques are aimed at protecting *artifacts*, such as digital data, image, and audio/video streams. This is very different from protecting *IP*: since a minor error can change the functionality of the IP and render the entire design useless, IP fingerprinting cannot be achieved in the same way.

Figure 4.1 depicts a simple symmetric scheme that achieves this. Each IP buyer provides the IP provider with his signature which is encrypted using the buyer's public key. The IP provider converts this encrypted message into fingerprinting constraints and integrates them with his watermarking constraints and the original design constraints. As a result, the synthesis tools will generate a piece of IP that has both the IP provider's watermark and the specific IP buyer's fingerprint. This allows IP provider to trace individual IP buyer (since each IP becomes unique with the buyer's fingerprint) and it also protects the buyer (since the buyer is the only entity who can interpret the signature via his secret

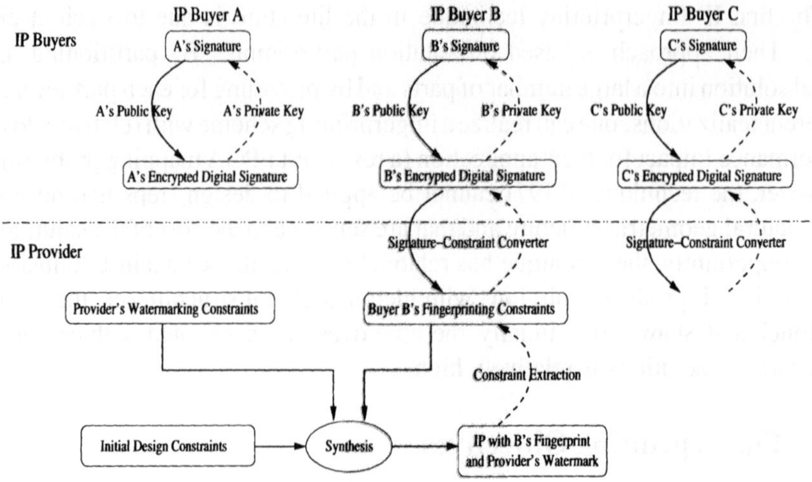

Figure 4.1. A symmetric interactive fingerprinting IP protection technique.

key). The provider can not resell this realization of the IP without the buyer's permission as it is customized to the buyer and carries his signature. This symmetric fingerprinting protection methodology relies on the fact that the IP provider fingerprints and delivers to each buyer a unique copy of functionally identical IP.

2.2 General Fingerprinting Assumptions

Fingerprints are characteristics of an object which are sufficient enough to distinguish it from other similar objects. Fingerprinting refers to the process of adding fingerprints to an object and recording them, or process of identifying and recording fingerprints that are already intrinsic to the object[162]. The core idea of fingerprinting is to give each user a copy of the object containing a unique fingerprint, which can be used to identify that user.

One of the most accepted model for fingerprinting[20, 18, 35, 162, 125] can be described as: In the original object, a set of marks is selected probabilistically, where a *mark* is one bit of information that has two slightly different versions. The distributor can choose one of the two versions of each mark to embed either a 0 or a 1 when the object is sold to a user, and thus construct a binary word which becomes the fingerprint of this user. Two general assumptions on the object to be fingerprinted are:

Error-tolerance assumption: the object should remain useful after introducing small errors or marks, and the user cannot detect the marks from the data redundancy. The more errors that the object can tolerate, the more places we can put these marks.

Marking assumption: two or more users may detect a few marks that differ in their copies, but they cannot change the undetected marks without rendering the object useless.

According to a taxonomy given by Wagner[162], the *statistical fingerprinting* is characterized as: given sufficiently many misused objects to examine, the distributor can gain any desired degree of confidence that he has correctly identified the compromised. The identification is, however, never certain. This is one of the fundamentals for many fingerprinting schemes[20, 18].

2.3 Context for Fingerprinting in IP Protection

Our goal is to protect IP through fingerprinting. The major difference between IPs and the objects mentioned in the previous section is that IPs are usually error-sensitive, which violates the error-tolerance assumption. However, one can see that this assumption's sole role is to guarantee a relatively large valid object space that can be easily generated[1]. Based on this observation, we propose the first requirement for the IP to be protected:

(1) The IP should be well-interpreted as a problem which has a large solution space. The sole role for the error-tolerance assumption is to guarantee a relatively large valid object space. Introducing errors is one way to create such space, but not the only way.

In the example in the footnote, the valid object space is trivially created compared to the possible huge cost of collecting the original n values (The only non-trivial part is to determine the delta value d_js.). It takes tremendous human and computer resources to design and implement a piece of IP, and we cannot afford to produce different copies by simply repeating the whole design process.

(2) The cost to derive the solution space should be negligible comparing to that of inventing the IP.

The last requirement, though not mandatory, is highly recommended for the sake of implementation:

(3) The existence of algorithms and/or state-of-the-art software which solves the problem. In our experience, these exist for many problems in the field of VLSI CAD. Furthermore, we require the fingerprinting protocols to be non-intrusive, i.e., the algorithm and/or the software will serve as a "blackbox".

2.4 Fingerprinting Objectives

A fingerprint, being the signature of the buyer, should satisfy all the requirements of any effective watermark:

- **High credibility.** The fingerprint should be readily detectable in proving legal ownership, and the probability of coincidence should be low.

- **Low overhead.** Once the demand for fingerprinted solutions exceeds the number of available good solutions, the solution quality will necessarily degrade. Nevertheless, we seek to minimize the impact of fingerprinting on the quality of the software or design.

- **Resilience.** The fingerprint should be difficult or impossible to remove without complete knowledge of the software or design.

- **Transparency.** The addition of fingerprints to software and designs should be completely transparent, so that fingerprinting can be used with existing design tools.

- **Part Protection.** Ideally, a good fingerprint should be distributed all over the software or design in order to identify the buyer from any part of it.

At the same time, the IPP business model implies that fingerprints have additional mandatory attributes:

- **Collusion-secure.** Different users will receive different copies of the solution with their own fingerprints embedded. These fingerprints should be embedded in such a way that it is not only difficult to remove them, but also difficult to forge a new fingerprint from existing ones (i.e., the fingerprinted solutions should be structurally diverse).

- **Runtime.** The (average) runtime for creating a fingerprinted solution should be much less than the runtime for solving the problem from scratch. The complexity of synthesis problem and the need for large quantity of fingerprinted solutions make it impractical to solve the problem from scratch for each individual buyer.

- **Preserving watermarks.** Fingerprinting should not diminish the strength of the author's watermark. Ideally, not only the fingerprinting constraints should not conflict with the watermarking constraints, any hint on the watermark from fingerprints should also be prevented as well.

From the above objectives, we extract the following key requirements for fingerprinting protocols:

- A fingerprinting protocol must be capable of generating solutions that are "far away" from each other. If solutions are too similar, it will be difficult for the seller to identify distinct buyers and it will be easy for dishonest buyers to collude. In most problems, there exist generally accepted definitions for distance or similarity between different solutions.

- A fingerprinting protocol should be non-intrusive to existing design optimization algorithms, so that it can be easily integrated with existing software tool flows.

- The cost of the fingerprinting protocol should be kept as low as possible. Ideally, it should be negligible compared to the original design effort.

3. Iterative Fingerprinting Techniques

We propose the iterative fingerprinting technique which can be applied to an arbitrary optimization/synthesis problem. It leverages the optimization effort already spent in obtaining a previous solution, yet generates a unique fingerprinted new solution. We develop specific fingerprinting approaches for four classes of VLSI CAD optimizations (graph coloring, partitioning, satisfiability, and standard-cell placement) to demonstrate this generic strategy.

3.1 Iterative Optimization Techniques

An instance of finite global optimization has a finite solution set S and a real-valued cost function $f : S \rightarrow \Re$. Without loss of generality, global optimization seeks a solution $s^* \in S$ which minimizes f, i.e., $f(s^*) \leq f(s)\ \forall s \in S$. This framework applies to most combinatorial domains (scheduling, coloring, partitioning, quadratic assignment, etc.); continuous optimizations can also be discretized to yield finite instances. Many optimization problems are NP-hard [61], and hence heuristic methods are often applied which use an iterative approach broadly described by the iterative global optimization template of Figure 4.2.

Typically, s' in Line 2 of Figure 4.2 is generated by a perturbation to s_i, i.e., $s' \in N(s_i)$ where $N(s_i)$ indicates the *neighborhood*, or set of all possible "neighbor" solutions, of s_i under a given neighborhood operator. Example operators include changing a vertex's color in graph coloring; swapping two cells in standard-cell placement; moving a vertex to a different partition in graph partitioning; etc. The collection of neighborhoods $N(s_i)$ implicitly defines a topology over S, which we denote as the *neighborhood structure*, N. Together with N, the cost function f defines a *cost surface* over the neighborhood topology, and iterative optimization searches this surface for (an approximation to) a globally minimum solution. Each iteration of Lines 2 through 4 is a *step* in the algorithm; the sequence of steps from step 0 until the algorithm terminates in Line 5 is a *run* of the iterative optimization algorithm.

We make two observations:

- Steps 2-4 of Figure 4.2 can be hierarchically applied to create very complicated metaheuristics. For example, the Kernighan-Lin [86] and Fiduccia-Mattheyses [53] graph partitioning heuristics are both greedy iterative opti-

Iterative Global Optimization	
1.	**for** $i = 0$ to $+\infty$
2.	Given the current solution s_i, generate a new trial solution s'
3.	Decide whether to set $s_{i+1} = s_i$ or $s_{i+1} = s'$
4.	**if** a stopping condition is satisfied
5.	**return** the best solution found

Figure 4.2. Basic template for iterative global optimization.

mizers with respect to a complicated *pass* move that is itself a move-based iterative optimization.[2]

- The complexity of the metaheuristic and its sensitivity to perturbations of the instance can be a vehicle for IPP: given a solution (say, an assignment of vertices to partitions) it is typically extraordinarily difficult to identify the instance (say, the weighted edges of a graph over the vertices) for which a given metaheuristic would return the solution.

3.2 Generic Approach

To maintain reasonable runtime while producing a large number of fingerprinted solutions, we will exploit the availability of iterative heuristics for difficult optimizations. Notably, we propose to apply such heuristics (i) in an *incremental* fashion, and (ii) to design optimization instances that have been perturbed according to a buyer's signature (or fingerprint). In the remainder of this section, we will focus on the creation of fingerprinted solutions and will not discuss the mechanics of encoding a buyer's plain text signature into a digital signature (normally as a pseudo-random bitstream), converting the pseudo-random bitstream into design constraints, and embedding the constraints into designs. Such techniques (using, for example, the cryptographic hash function MD5, the public-key cryptosystem RSA, and the stream cipher RC4) have been discussed at length in the recent literature on IP protection (e.g., [81]).

Generic Iterative Fingerprinting Approach	
1.	Create the watermarked instance I_0 with provider's watermark embedded into the initial instance I;
2.	Generate a watermarked initial solution S_0 from I_0 using a *from-scratch* optimization;
3.	**for** $j = 1$ to n (n is the number of buyers)
4.	Create a fingerprinted instance I_j with the j-th buyer's fingerprint F_j added into I_0;
5.	Starting from S_0 as the initial solution, apply an *incremental* optimization to generate a new fingerprinted solution S_j for I_j;

Figure 4.3. The generic iterative approach for generating fingerprinted solutions.

Figures 4.3 and 4.4 outline the basic approach. Given a design instance I, our approach starts by embedding the provider's watermark into I and generating an initial watermarked solution S_0 using an (iterative) optimization heuristic in "from-scratch" mode. This can be achieved by any of the constraint-based watermarking techniques reported in literature. Then we use this solution S_0 as the "seed" to create fingerprinted solutions as follows. For a given buyer, we embed the buyer's signature into the design as a fingerprint (e.g., by perturbing the weights of edges in a weighted graph), which yields a fingerprinted instance I_j. Instead of solving I_j "from-scratch", we start with S_0 as the initial solution and perform an *incremental* iterative optimization step to obtain solution S_j.

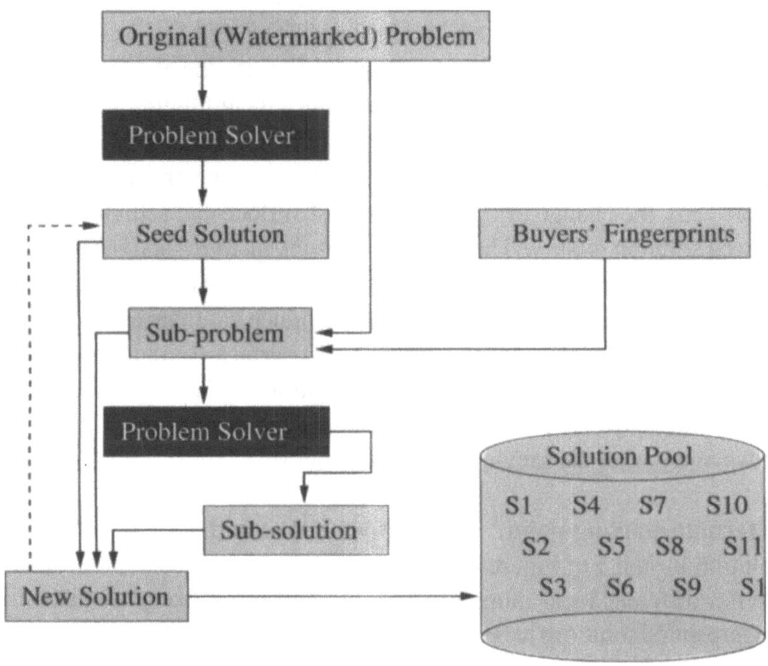

Figure 4.4. Iterative fingerprinting technique in the system design process.

This fingerprinting approach, compared to the naïve one in Figure 4.1, has the following advantages:

- **Shortened runtime.** We observe that the iterative optimization heuristic will be applied using a known high-quality solution S_0 as the starting point, so the runtime until the stopping criterion is reached (e.g., arriving at a local minimum) will be much less than that of a from-scratch optimization. Essentially, we leverage the design optimization effort that is inherent in the "seed" solution S_0.

- **Distinct solutions.** The starting point S_0, as the solution with only watermark-related additional constraints, should be a (good) local minimum. Therefore, it is relatively difficult for the iterative optimizer to get over such local minimum if we use S_0 as the starting point. However, the addition of fingerprinting constraints will subtly changes the problem instance and hence the optimization cost surface. Such change may affect the local minimality of S_0 and help the iterative optimizer to find a new local minimum, the fingerprinted solution.

- **Additional fingerprint.** The change of optimization cost surface not only prevents the iterative optimizer from falling into the same local minima as before, it will also direct the iterative heuristic to a new local minimum following a rather unique path and further fingerprints the design. As noted above, it is exceedingly difficult to reverse-engineer the particular weighting of the instance for which a given solution is a local minimum[3].

- **Improved solution quality.** As noted in the metaheuristics literature [121, 155], the change of optimization cost surface can actually lead to improved solution quality. The method for *problem-space* and *heuristic-space* development perturb a given instance to allow a given optimization heuristic to escape local minima. The perturbations induce alternate cost surfaces that one hopes are correlated to the original cost surface (so that good solutions in the new surface correspond to good solutions in the original), yet which have sufficiently different structure (so that the optimization heuristic can move away from the previous local minimum).

- **Alternate starting point.** Alternatively, we could use S_{j-1} as the initial solution in step 5 of Figure 4.3. Then every fingerprinted solution will start from a different local minimum and this will more likely to make all the fingerprinted solutions to be "far away" from each other. The ultimate benefit is to reduce the change of collusion. However, all the previous fingerprinting constraints are inherent in S_{j-1} and this may make the fingerprinted instance over-constrained.

3.3 VLSI Design Applications

In this section , we develop specific fingerprinting approaches for four classes of VLSI CAD problems. We first discuss two classic examples for iterative optimization algorithms: *partitioning and standard-cell placement.* Then we explain how iterative fingerprinting approach can be applied to other optimization problems, for example the *graph coloring* (GC) problem, which may not be solved by iterative improvement. Finally, we claim that this approach is also applicable to decision problems such as *Boolean satisfiability* (SAT) problem.

3.3.1 Partitioning

Given a hyperedge- and vertex-weighted hypergraph $H = (V, E)$, a k-*way partitioning* of V assigns the vertices to k disjoint nonempty partitions. The k-*way partitioning problem* seeks to minimize a a giver objective function $c(P^K)$ with partitioning as its parameter. A standard objective function is *cut size*, i.e., the number of hyperedges whose vertices are not all in a single partition. Constraints are typically imposed on the partitioning solution, and make the problem difficult. For example, the total vertex weight in each partition may be limited (*balance constraints*), which results in an NP-hard formulation [61]. To achieve flexibility and speed in addressing various formulations, move-based iterative optimization heuristics are typically used, notably the Fiduccia-Mattheyses (FM) heuristic [53]. In our partitioning testbed, we use the recent CLIP FM variant [50] and the net cut cost function.

For a given partitioning instance I_0, we iteratively construct a sequence of fingerprinted solutions according to the following steps.

1 Generate an initial partitioning solution S_0 by finding the best solution out of 40 starts of CLIP FM for instance I_0.
2 Reset all hyperedge weights to 20.
3 According to the jth user's fingerprint, select a subset $E' \subset E$ of size equal to some percentage of the total number of hyperedges in H, and increment the weight of each hyperedge $e \in E'$ by +/- 19 (also according to the user's fingerprint). This yields instance I_j.
4 Partition the hypergraph instance I_j using a single start of CLIP FM, using S_0 (the initial non-fingerprinted solution) as the starting solution.[4] This yields the fingerprinted solution S_j.
5 If another fingerprinted solution is needed, return to Step 2.

3.3.2 Standard-Cell Placement

The standard-cell placement problem seeks to place each cell of a gate-level netlist onto a legal site, such that no two cells overlap and the wirelength of the interconnections is minimized. We iteratively construct a sequence of fingerprinted placement solutions according to the following steps (note that our approach is compatible with the LEF/DEF and Cadence QPlace based constraint-based watermarking flow presented in [81]).

1 Given an instance I_0 in LEF/DEF format, apply the placer (Cadence QPlace version 4.1.34) to generate an initial placement solution S_0.
2 Reset the weights of all signal nets to 1.
3 According to the jth user's fingerprint, select a subset $N' \subset N$ of the signal nets in the design, and set the weight of each net in N' to 10. This yields a fingerprinted instance I_j.

4 Incrementally re-place the design, starting from the current solution S_{j-1} and using the new net weighting. This is achieved by invoking the *Incremental Mode* of the *QPlace* tool, and yields the fingerprinted placement solution S_j.
5 Save the new placement solution S_j as the current solution.
6 If another fingerprinted solution is needed, return to Step 2.

3.3.3 Graph Coloring

The graph vertex coloring (**GC**) optimization seeks to color a undirected graph with as few number of colors as possible, such that no two adjacent vertices receive the same color. GC has a lot of applications in real life, for example, the register allocation problem, the cache-line coloring problem, wavelength assignment in optical networks, and channel assignment in cellular systems. There exist well-established GC benchmark graphs and algorithms [174, 175]. The GC algorithms can be classified into three categories: exact [42], constructive [69], and iterative improvement [55, 77]. It has been shown that iterative improvement methods (such as simulated annealing and generic tabu search), to which we can easily apply the iterative fingerprinting approach discussed above, are the most effective, in particular for random graphs. However, Coudert finds that exact coloring for real-life CAD-related graphs is easy [42]. It becomes important and interesting to study whether the proposed iterative fingerprinting technique is applicable when the underlying optimization algorithm does not possess the iterative improvement nature.

Given a graph $G(V, E)$ and an algorithm $\mathcal{A}$ (not necessary to be iterative improvement method), we iteratively construct a sequence of fingerprinted coloring solutions as follows:

1 Obtain a coloring solution S_0 by applying algorithm $\mathcal{A}$ to graph G: $\{C_1, \ldots, C_k\}$, where each C_i is an independent set and receives exact one color;
2 Select $\{C_{i_1}, \ldots, C_{i_l}\} \subset \{C_1, \ldots, C_k\}$ according to the j-th user's fingerprint;
3 Create the fingerprinted graph $G'(V', E')$

3.0 $V' = V; E' = E;$
3.1 for ($j = 1, \cdots, l$)
3.2 { if (C_{i_j} is a maximal independent set)
3.3 $V' = V' \setminus C_{i_j}$; $E' = E' \setminus \{(u, v) | u \in C_{i_j}$ or $v \in C_{i_j}\}$;
3.4 else
3.5 { select $v^j \in C_{i_j}$ randomly; $V' = V' \setminus C_{i_j} \cup \{v^j\}$;
3.6 $E' = E' \setminus \{(u, v) | u \in C_{i_j}$ or $v \in C_{i_j}\} \cup \{(u, v^j) | u \notin C_{i_j}$ and $\exists v \in C_{i_j}, s.t. (u, v) \in E\}$;
3.7 }
3.8 }
3.9 G' = watermarking (G', user's fingerprinting constraints);

4 Obtain a coloring solution for graph G';
5 Create the fingerprinted solution for graph G

5.1 for ($j = 1, \cdots, l$)

5.2 { if (C_{i_j} is a maximal independent set)
5.3 assign all vertices in C_{i_j} a new color;
5.4 else
5.5 assign all vertices in C_{i_j} the same color of v^j;
5.6 }

6 Go to step 2 if another fingerprinted solution is needed;

A coloring solution is essentially a partition of vertices into disjoint independent sets (IS) where all vertices in the same IS will be assigned one color. We start from a (watermarked) solution S_0 and select part of it (l ISs) to embed fingerprint in step 2. The selection of these ISs could be based on the user's fingerprint, but the majority of the fingerprint will be embedded into the fingerprinted graph $G'(V', E')$ in step 3. We treat the selected ISs differently according to their maximality because there may still be possible to include more vertices to a non-maximal IS. We preserve all the maximal ISs (MISs) in the selection by deleting them from the graph (steps 3.2 and 3.3). We also preserve each of the rest non-maximal IS, C_{i_j}, by collapsing it into one single node v^j and connect v^j to all the vertices that are neighbors of any vertices in C_{i_j} (steps 3.4-3.6) and keeping it in the new graph to keep the chance of improving such IS alive. Finally we apply any of the existing GC watermarking schemes [132] to embed the user's fingerprinting constraint and form the fingerprinted graph G' in step 3.9.

The fingerprinted graph G' will have smaller size than the original graph G and hence the run time of finding a good coloring solution will be less than that of coloring graph G. However, the solution that we obtain in step 4 will be one for G' which has different vertices from the desired graph G. In step 5, we convert it to a fingerprinted coloring solution to G by giving each missing MISs a new color and all vertices in other missing ISs, C_{i_j}, the same color as the one received by their collapsed representative v^j in G'.

We first mention that the algorithm $\mathcal{A}$ can be any graph coloring algorithm, not necessarily to follow the iterative improvement approach. Secondly, the fingerprinted solution we obtain in step 5 will be different from the initial solution S_0 in step 1. This is because that S_0 does not guarantees the satisfaction to the user's fingerprinting constraints, which have been added in step 3.9 to the graph. For any good watermarking technique, the coincidence that a random solution meeting the watermarking constraints should be extremely low [132]. Finally, we have already explained that the run time to get the fingerprinted solution should be less than that of solving the problem from scratch because we are coloring a smaller graph. Another reason for the shortened run time is that we are reusing the efforts that we put to find the intnial solution S_0 by preserving the (selected) independent sets, which are presumably good quality ISs.

3.3.4 Satisfiability

As the final example, we show that the proposed iterative fingerprinting approach is not limited to optimization problems only by studying the boolean satisfiability (SAT) problem, the most representative NP-complete decision problem. The SAT problem seeks to decide, for a given formula, whether there exists a truth assignment for the variables that makes the formula true. Because of its discrete nature, SAT appears in many contexts in the field of VLSI CAD, such as automatic pattern generation, logic verification, timing analysis, delay fault testing and channel routing. A brief survey on SAT and its application in EDA can be found in [109]. We necessarily assume that the given SAT instance is satisfiable and that it has sufficient large solution space to accommodate multiple fingerprinted solutions.

Given a formula $\mathcal{F}$ on a set of boolean variables $V = \{v_1, \cdots, v_n\}$, we iteratively construct a sequence of fingerprinted solutions according to the following steps.

1 Solve $\mathcal{F}$ for an initial solution $S_0 : \{v_1 = b_1, \cdots, v_n = b_n\}$, where $b_i \in \{0, 1, -\}$;

2 According to the jth user's fingerprint, select a subset of variables: $V' = \{v_{i_1}, \cdots, v_{i_l}\} \subset V$;

3 Create the fingerprinted formula $\mathcal{F}'$

3.0 $\mathcal{F}' = \mathcal{F}$;
3.1 for ($j = 1, \cdots, l$)
3.2 { if ($b_{i_j} = 1$)
3.3 $\mathcal{F}' = \mathcal{F}'_{v_{i_j}}$; /* $\mathcal{F}'_{v_{i_j}}$ is the *cofactor* of $\mathcal{F}'$ with respect to variable v_{i_j}. */
3.4 else if ($b_{i_j} = 0$)
3.5 $\mathcal{F}' = \mathcal{F}'_{v'_{i_j}}$; /* $\mathcal{F}'_{v'_{i_j}}$ is the *cofactor* of $\mathcal{F}'$ with respect to variable v'_{i_j}. */
3.6 else $\mathcal{F}' = \mathcal{F}' \setminus v_{i_j}$; /* $\mathcal{F}' \setminus v_{i_j}$ means removing both v_{i_j} and v'_{i_j} from $\mathcal{F}'$. */
3.7 }
3.8 $\mathcal{F}' =$ watermarking ($\mathcal{F}'$, user's fingerprinting constraints);

4 Solve $\mathcal{F}'$ and get an assignment to all the variables in $V - V'$;

5 Create the fingerprinted solution for formula $\mathcal{F}$

5.1 for ($j = 1, \cdots, l$)
5.2 $v_{i_j} = b_{i_j}$;

6 Go to step 2 if another fingerprinted solution is needed;

For a satisfiable formula $\mathcal{F}$, a solution is an assignment of 0 (false), 1 (true), or - (don't care) to each of the variables[6]. We fix the assignment to a selected subset of variables in the initial solution (step 2) and build a fingerprinted formula $\mathcal{F}'$ on the rest of the variables (step 3). We first simplify the formula by considering the fixed values of selected variables v_{i_j}. If v_{i_j} is assigned *true*, all clauses with v_{i_j} are satisfied automatically and we can also safely remove v'_{i_j} from the formula. As a result, we get the *cofactor* [7] of $\mathcal{F}'$ with respect to variable v_{i_j} (step 3.3). It is similar to the case when v_{i_j} is assigned *false* (step 3.5). However, if v_{i_j} is assigned *don't care* in a solution, which means that the value of v_{i_j} will not affect the satisfaction to the formula, then we can safely remove both v_{i_j} and v'_{i_j} from the formula (step 3.6). In the last step 3.8, we apply any of the existing SAT watermarking techniques [133] to add user's fingerprinting constraints into the formula $\mathcal{F}'$. Then we solve the fingerprinted formula $\mathcal{F}'$ (step 4) and combine the result with the values of the selected variables v_{i_j} to form the fingerprinted solution (step 5).

Unlike the optimization problems, such as partitioning and GC, where the quality of the solution is crucial, the effectiveness of SAT fingerprinting techniques is measured by the run-time and distinctness among fingerprinted solutions. We will give quantitative analysis for both in the experimental results section. Here we only mention that the reduction on run time is a result of 1) the cofactoration in steps 3.3 and 3.5 as well as in 3.6 which reduce the size of the (fingerprinted) SAT instance and 2) the preservation of the values for a selected subset of variables which keeps the effort in finding the initial solution.

3.4 Experimental Results

We have conducted experiments on benchmark data for the above four problems. The goal is to verify that the proposed iterative fingerprinting approach meets the fingerprinting objectives and requirements as we discussed earlier in section 2. In particular, we focus our analysis on 1) the run time for creating multiple fingerprinted solutions, 2) the quality of the fingerprinted solutions (except the SAT problem), and 3) the distinctness among the fingerprinted solutions. We further make the following notifications:

- Robustness of the fingerprint. It is important to have robust fingerprints. However, it is not our intention to propose any robust fingerprinting methods and this paper does not make any contribution on it either. In light of the fact that fingerprint can also be viewed as the user's watermark, we apply the existing watermarking techniques to embed fingerprint and rely on these techniques to provide the robustness.

- Non-intrusive to existing CAD tools. In the proposed iterative fingerprinting approach, we require only the input/output interface of the CAD tool (partitioner, placer, or any GC and SAT solver). We create fingerprinted problem instances based on the solution provided by the tools and feed it into the tools again. Throughout the fingerprinting process, the tools can be viewed as a "black box".

- Tool independent. Clearly from the pseudo-codes in section 3.3. we see that the proposed iterative fingerprinting approach does not dependent a specific algorithm or CAD tool. We emphasize here that it is not our goal to compare the performance of different tools for the same problem. Instead, we will demonstrate the run time saving of the iterative fingerprinting approach over solving the instance from scratch.

- Watermark preservation. It is required that the user's fingerprint should not offend author's watermark by either violating the watermarking constraints or leaking some information about the watermark. Notice that in our approach, the watermarking constraints are kept during the fingerprint process hence they are considered as "original" design constraints and will be satisfied by all fingerprinted solutions and be ready to be revealed to establish that author's authorship. Furthermore, the methods for fingerprinting constraints generation and embedding can be independent of ones used for watermarking constraints, therefore one should not get any hint about the watermark from a fingerprinted solution.

Partitioning

We test our fingerprinting method on 3 standard test cases from the ISPD-98 Benchmark Suite [4] [3]. These correspond to internal IBM designs that have been recently released to the VLSI CAD community. We apply the CLIP FM partitioner with a 10% balance constraint, and the actual vertex weights. For each test case, a single experimental trial generates an initial solution, followed by a sequence of 20 fingerprinted solutions (i.e., we go through Step 2 of the method in Section 4.1 a total of 20 times). Table 4.2 reports the average results of 20 independent trials.[8] We report the maximum and average solution cost for the initial solutions S_0, as well as the maximum and average solution costs for the fingerprinted solutions S_i. We also report the maximum and average CPU times required to generate an initial solution S_0 or a fingerprinted solution S_i. (All CPU times that we report are for a 300MHz Sun Ultra-10 running Solaris 2.6.) Finally, we report the minimum and average Hamming distances (i.e., number of transpositions required to transform one solution into another) over all $C(21, 2)$ pairs among the solutions $S_0, S_1, \ldots, S_{20}$. The data show that the fingerprinted solutions: (i) require much less CPU to generate than the original solutions (by factors ranging from 18 to 77); (ii) are reasonably distinct from each other and from the original solutions; and (iii) can even have better average quality than the original solutions (which we attribute to the similarity between our fingerprinting methodology and the problem-space iterative optimization metaheuristic [121]).

Test Case:	IBM01	IBM02	IBM03
Number of Vertices	12752	19601	23136
Number of Hyperedges	14111	19584	27401

Table 4.1. Test cases for partitioning experiments.

Test Cases	S_0 Cost		S_0 CPU Time		S_i Cost		S_i CPU Time		Distance	
	Max	Ave	Max	Ave	Max	Ave	Max	Ave	Min	Ave
IBM01	308	252.2	261	187.6	307.2	253.8	3.25	2.42	16.3	71.1
IBM02	296	272.0	379	329.8	278.5	273.1	15.2	8.4	7.3	41.2
IBM03	1047	881.5	808	695.5	912.5	867.5	36.1	18.4	66.6	263.4

Table 4.2. Results for the fingerprinting flow on three standard bipartitioning test cases. Tests were run using actual cell areas, and a partition area balance tolerance of 10%. Each trial consists of generating an initial solution, then generating a sequence of 20 fingerprinted solutions. All results are averages over 20 independent trials.

Standard-Cell Placement

For standard-cell placement, we have applied our fingerprinting technique to the four industry designs listed in Table 4.3. For each test case, we generate an initial solution S_0 and a sequence of 20 different fingerprinted solutions $S_1, \ldots, S_{20}$; for each fingerprinted solution, the previous fingerprinted solution is used as the initial solution for QPlace Incremental Mode. Table 4.4 presents a detailed analysis of the solutions obtained for the Test2 instance. We measure the structural difference between solutions as "Manhattan Distance": the sum over all cells in the design of the Manhattan distance between the two placed locations for each cell. We see that a fingerprint that perturbs just 1% of the net weights achieves reasonably large Manhattan distance from S_0, and that the incremental optimization saves a significant amount of CPU versus the from-scratch optimization. Again, there is a "problem-space metaheuristic" effect in that the fingerprinted solutions are typically of higher quality than the original solution. A summary of results for all four test cases is given in Table 4.5. From this table we can see that we can reduce the time to generate the next fingerprinted solution while maintaining the quality as well as producing a unique solution.

Test Case:	Test1	Test2	Test3	Test4
Number of Cells	3286	12133	12857	20577
Number of Nets	2902	11828	10880	25634

Table 4.3. Test cases for standard-cell placement experiment.

Graph Coloring

We have implemented our proposed GC fingerprinting technique and applied it to real-life benchmarks and the DIMACS challenge graph. The real-life

Solution	1% Nets Weighted			2% Nets Weighted			5% Nets Weighted		
	Cost	Dist.	CPU	Cost	Dist.	CPU	Cost	Dist.	CPU
Orig(S0)	1.000	0	6:24	1.000	0	6:24	1.000	0	6:24
S1	0.994	3.532	2:58	1.001	3.757	2:58	1.020	3.946	2:58
S2	0.986	4.341	2:57	0.994	3.858	2:58	1.023	4.518	2:54
S3	0.981	4.666	2:58	0.997	4.032	3:03	1.016	4.536	2:51
S4	0.982	4.545	3:02	0.997	4.203	3:03	1.018	4.990	2:52
S5	0.979	4.807	3:01	0.986	4.574	2:57	1.005	5.095	2:54
S6	0.981	4.989	2:59	0.990	5.115	3:01	1.007	5.272	2:55
S7	0.971	5.121	3:30	0.982	5.359	3:33	1.014	5.543	3:28
S8	0.972	5.257	3:32	0.989	5.510	3:30	1.005	5.657	3:28
S9	0.970	5.513	3:38	0.979	5.804	3:32	1.004	5.972	3:29
S10	0.970	5.820	3:33	0.976	6.354	3:29	0.998	6.089	3:29
S11	0.970	6.011	3:34	0.977	6.253	3:30	1.005	6.248	3:30
S12	0.974	6.030	3:34	0.983	6.794	3:34	1.006	6.264	3:29
S13	0.974	6.738	3:35	0.974	6.554	3:34	0.998	6.356	3:30
S14	0.972	6.826	3:38	0.966	6.441	3:32	1.002	6.742	3:30
S15	0.978	6.975	3:36	0.972	6.354	3:32	1.004	7.106	3:28
S16	0.976	6.762	3:39	0.976	6.107	3:33	0.992	7.034	3:29
S17	0.978	6.911	3:38	0.975	6.654	3:34	0.992	7.129	3:27
S18	0.970	7.144	3:39	0.974	6.657	3:34	0.990	7.198	3:26
S19	0.974	7.254	3:37	0.969	6.440	3:28	0.995	7.161	3:27
S20	0.969	7.331	3:40	0.977	6.414	3:34	0.986	6.989	3:26

Table 4.4. Standard-cell placement fingerprinting results for the Test2 instance. We report CPU time (mm:ss) needed to generate each solution, as well as total wirelength costs normalized to the cost of the initial solution S_0. Manhattan distances from S_0 are given in 10^6 microns.

benchmark graphs are converted from register allocation problem of variables in real codes with known optimal solutions [175]. They are easy to color and almost all the original and fingerprinted graphs are colored instantaneously with no extra colors. However, the DIMACS challenge graph, which is a random graph with 1000 vertices and an edge probability slightly larger than 0.5, is hard and the optimal solution is still open [174]. We report our results on the latter as further evidence to the tradeoff between solution quality and fingerprint's credibility. We also show the run time saving for generating new solutions by the iterative fingerprinting technique.

We first color the graph once and obtain an 86-color "seed" solution. Then we choose different percentage of independent sets (ISs) to create fingerprinted graphs by preserving the selected independent sets as discussed in section 3.3. We use the watermarking method called "adding edges" reported in [132] to embed a set of fingerprinting constraints, which is a pseudo-random bitstream of the same length as the number of vertices in the new graph. We color

Test cases	% weighted nets	Normalized Cost		CPU (sec)		Manhattan Dist.	
		Max	Ave	Max	Ave	Min	Ave
Test1	Original	1.000	1.000	97	97	0	0
	1%	1.009	0.986	53	51	0.530	0.861
	2%	1.003	0.987	54	50	0.509	0.886
	5%	1.024	0.992	58	53	0.801	1.084
Test2	Original	1.000	1.000	384	384	0	0
	1%	0.994	0.976	220	205	3.532	5.829
	2%	1.001	0.982	214	202	3.757	5.662
	5%	1.023	1.004	210	198	3.946	5.992
Test3	Original	1.000	1.000	344	344	0	0
	1%	1.000	0.988	168	164	3.232	4.048
	2%	1.007	0.995	165	162	3.239	4.123
	5%	1.015	1.009	166	163	3.503	4.423
Test4	Original	1.000	1.000	960	960	0	0
	1%	1.006	0.983	510	499	11.39	15.53
	2%	0.997	0.987	506	485	13.06	17.44
	5%	1.031	1.018	514	496	12.35	18.02

Table 4.5. Summary of results for fingerprinting of all four standard-cell placement instances. "Original" lines refer to the initial solutions S_0. All other lines refer to fingerprinted solutions $S_i, i > 0$. Manhattan distance is again expressed in 10^6 microns.

	Nodes	Edges	Edge Prob.	solution Ave.	solution Best	Run-time (hours)
Original Graph	1000	249826	0.5002	86	86	15.45
Recolor 20% ISs	222	13630	0.5556	86	86	1.10
Recolor 30% ISs	319	26908	0.5305	86	86	1.46
Recolor 40% ISs	397	40712	0.5179	86	86	2.02
Recolor 50% ISs	495	62726	0.5130	87	87	3.07
Recolor 60% ISs	593	89006	0.5071	87	87	4.55
Recolor 70% ISs	694	121591	0.5056	87.2	87	6.42

Table 4.6. Results for coloring the DIMACS challenge graph with iterative fingerprinting.

each fingerprinted graph five times. Parameters of the fingerprinted graphs and solutions, along with the average runtime, are reported in Table 4.6. The first column gives the percentage of independent sets that we decide to recolor, the rest ISs will be preserved; the second column is the number of vertices in the fingerprinted graphs, which is the total of vertices in the recolored ISs and the number of preserved non-maximal ISs; the edges in the third column include those added as fingerprinting constraints, which is the same as the number of vertices; the next two columns show the average number of colors we need to color a fingerprinted graph and the best coloring we have in five tries; the last column is the average run time to find one solution.

We can see that as the number of recolored ISs goes up from 20% to 70%, the fingerprinted graph will have more vertices to accommodate more fingerprinting constraints. This consequently increases the credibility of the fingerprint. However, the quality of the solution, in terms of the number of colors used to color the graph, degrades despite more time is spent to find a solution. The degradation of solution quality is the direct result of adding more fingerprinting constraints. The longer run time is due to the fact that the size of the fingerprinted graph becomes larger and more structural information from the "seed" solution is removed as we are recoloring more ISs. Still, we see significant run time savings over the original from-scratch run time (15+ hours) in all cases.

Satisfiability

The SAT instances in our experiments, which are generated from the problem of inferring the logic in an 8-input, 1-output "blackbox", are from DIMACS [174]. All instances that we use are satisfiable and WalkSAT [173] is used as the satisfiability solver. As described in early sections, we begin by solving each instance initially to obtain the "seed" solution. The approach fixes and therefore preserves a subset of the initial solution according to the user's fingerprint. To simplify this procedure, in our experiments, k% of the variables are randomly selected to be preserved from the initial solution. Once these variables have been selected and preserved, they are removed from the instance, leaving a simplified instance. Specifically, the instance is simplified by removing all clauses which are satisfied by the preserved variables, and all complemented versions of the preserved variables are removed form the instance. We find a solution for the smaller fingerprinted instances and compare them to the original solution. The solution is representative of the solutions which each user would receive after embedding their signature. We compare the Hamming distance of the obtained solution to the original solution in order to determine the credibility of the solution and the approach. The distance of two solutions $\mathcal{S} = \{s_1, s_2, \ldots, s_n\}$ and $\mathcal{T} = \{t_1, t_2, \ldots, t_n\}$ is defined as: $dist(\mathcal{S}, \mathcal{T}) = \sum_{i=1}^{n} |s_i - t_i|$.

Table 4.7 reports the results when we maintain 20%, 30%, and 50% of the "seed" S_0 solution. From the last two rows, we can see that on average, we are able to achieve solutions which are around 20% different from the seed with a near 40% CPU time savings. At first sight, one may expect that the more variables we preserve, we will have significant reduction in runtime, since the new instance is smaller. Furthermore, one may expect that the distance between the new solution and the seed solution is smaller and that therefore the solution is less credible. Interestingly, if the experimental results are analyzed statistically this is not the case. The CPU savings of each of the 20%, 30% and 50% cases is essentially the same. The explanation is the following. Although the size of the instances shown in the table decrease with the percentage of the variables preserved, the structural difficulty of these instances increases. The important observation is that the original instances of the problem have nigh numbers of solutions. The difficulty of the instances increases due to the fact that variables are preserved at random and many initially feasible solutions become infeasible. Therefore, solvers have a difficult time more thoroughly traversing the solution space, and the additional savings in terms of CPU time is minimal. On a positive note, the means that a large portion of the solution can be preserved with very little overhead.

Original Solution			20% Preserved			30% Preserved			50% Preserved		
File	Var.	CPU	Var.	Dist.	CPU	Var.	Dist.	CPU	Var.	Dist.	CPU
ii8a1	66	8.0	52	20	6.6	46	16	7.2	33	21	6.8
ii8a2	180	9.2	144	35	7.0	126	33	6.8	90	24	7.0
ii8a3	264	9.4	211	48	8.4	184	43	7.4	132	38	7.4
ii8a4	396	12.8	316	78	8.4	277	60	7.8	198	56	9.2
ii8b1	336	9.6	268	80	9.0	235	53	7.4	168	55	8.4
ii8b2	576	13.8	460	146	7.6	403	138	7.2	288	147	7.0
ii8b3	816	18.8	652	203	7.8	571	198	8.4	408	224	8.0
ii8b4	1068	30.8	854	272	8.2	747	246	9.0	534	277	7.4
ii8c1	510	12.4	408	92	8.0	357	103	8.4	255	72	7.8
ii8c2	950	17.4	760	218	8.6	665	238	7.8	475	246	7.4
ii8d1	530	12.2	424	68	8.6	371	60	8.8	265	71	7.8
ii8d2	930	17.2	744	246	7.8	651	251	8.0	465	212	7.2
ii8e1	520	12.0	416	82	8.4	364	56	8.6	260	68	6.8
ii8e2	870	17.2	696	165	7.4	609	223	8.0	435	98	8.2
Ave. Distance (%)			-	22%	-	-	20%	-	-	19%	-
Ave. CPU Saving (%)			-	-	38%	-	-	39%	-	-	41%

Table 4.7. Number of undetermined variables (*Var.*), average distance from original solution (*Distance*), and average CPU time (in $1/100^{th}$s of a second) for fingerprinting SAT benchmarks.

4. Constraint-Based Fingerprinting Techniques

Now we further address the following fingerprinting problem: How to generate a large number of high quality solution for a given optimization problem by solving the initial problem only once. We propose a general technique which enables fingerprinting at all level of design process and is applicable to an arbitrary optimization step. In addition we also discuss how to select a subset of k solutions from the pool of n solutions so that the solutions are maximally different.

The key idea is to superimpose additional constraints on the problem formulation so to guarantee that the final solution can be in a straightforward way translated into k different high quality solution. In order to make our discussion concrete we focus on a single NP-complete problem - graph coloring. We tested the new fingerprinting on a number of standard benchmarks. Interestingly, while on random graphs it is relatively difficult to produce a large number of solutions without nontrivial quality degradation, on all real-life compilation graphs we are able to generate millions of solution which are all optimal.

4.1 Motivation, New Approach, and Contributions

The partitioning based FPGA fingerprinting technique[97] partitions the problem into a set of subproblems, and introduces constraints to connect these small problems if necessary, then solves each subproblem independently. This method has very poor performance unless the original problem has specific structure. The iterative approach we discuss above solves the problem once, then generates a relatively small problem based on this solution. Re-solving the small problem will give us possibly new solutions. Cost for solving small instance is usually much lower than is for the original, but when the request for different solutions are huge, this overhead cannot be ignored and moreover, different solutions are not guaranteed[9]. Is it possible to cut the runtime even further while generating guaranteed different solutions?

Moreover, with all the solutions scattered and unorganized, we need to keep a huge database to recorder the one-to-one map between IPs and individual IP buyers. Maintaining such database may be costly. Imagine that we have a 10,000 copies of coloring solutions to a 1,000-node graph, for each solution, we have to remember the color assigned to each node. This requires at least 10MB storage if we use one byte to represent colors, leaving alone the cost for keeping IP buyer's (encrypted) signature files. Can we have the solutions well-organized and easy-to-maintain?

We now present a fingerprinting technique to overcome these difficulties illustrated by the graph coloring problem. Figure 4.5 shows the generic approach of the new methodology. It consists of two phases, first we develop methods for generating many GC solutions with the smallest overhead, then we provide scheme to distribute these solutions among potential users.

This approach is highlighted by the solution generation phase, in which we first add fingerprinting constraints to the original (or watermarked) problem; a set of solution generation rules are created at the same time; after calling the problem solver once and get one seed solution, we can apply the solution generation rules and instantaneously create plenty solutions, each is different from another. This new approach provides six main benefits:

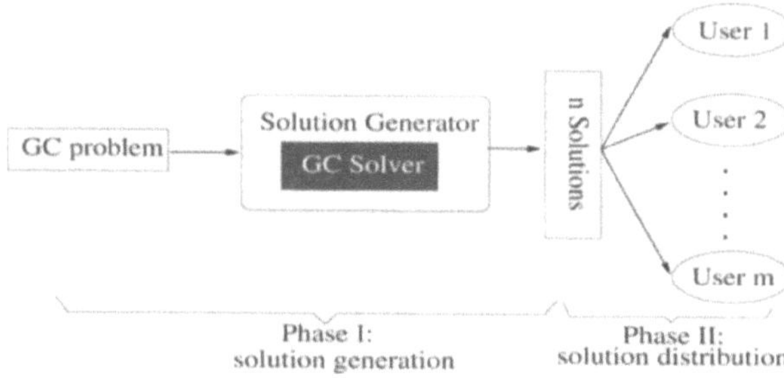

Figure 4.5. Two-phase fingerprinting technique for IP protection: generating n solutions and distributing among m users.

1 Since we call the solver only once, the run-time overhead for generating many solutions over that for one single solution is almost zero.

2 In three of the four techniques that we have implemented, the number of solutions can be controlled and the solutions are guaranteed distinct.

3 The actual solutions are not important, we can retrieve all the solutions from the seed solution and the solution generation rule.

4 The IP provider's signature can be embedded in the fingerprinting process without additional watermarking techniques.

5 Both symmetric and asymmetric fingerprints can be created by this method.

6 With proper distribution schemes, the techniques can be collusion-secure.

4.2 Generic Constraint-Addition IP Fingerprinting

As we have shown in Figure 4.5, the constraint-addition fingerprinting procedure consists of two phases: the solution generation phase and the solution distribution phase. Figure 4.6 depicts the flow of solution generation.

We start from the original problem (or alternatively the one with IP provider's watermarks), an augmented problem is build by adding the fingerprinting constraints. This step can be combined with the author's signature embedding. Basically, we can select the fingerprinting constraints based on, and thus hide, the author's watermark. Then, associated with this fingerprinted problem, we get a set of (simple) rules telling us how to create various solutions to the original problem from one to this augmented problem. For example, a trivial method for SAT could be, to put constraints such that a subset of variables become "don't-care". The solution generation rules, in this case, may read: altering the values assigned to the following variables to create solution. Next, we call the problem solver to solve this augmented problem. For each solution we get

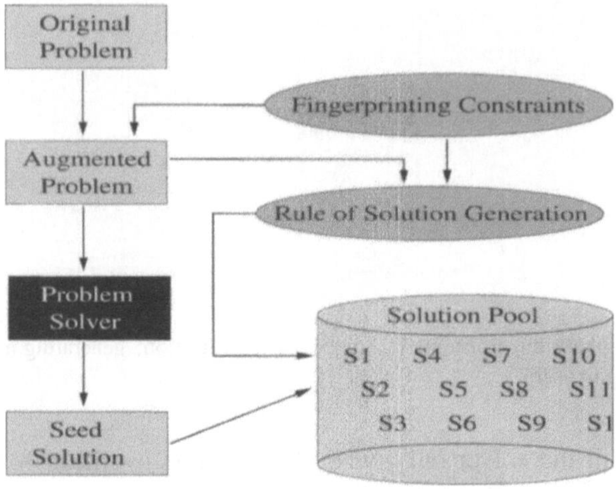

Figure 4.6. Solution generation phase of the constraint addition based fingerprinting technique in the system design process.

from the solver, we are able to apply the solution generation rules and build a pool of valid solutions.

Since the solutions are built around the seed solution according to the solution generation rules, the complete information of any solution can be retrieved from the seed and the parameters when applying the rules. It is not necessary to store the entire pool of solutions. Moreover, in the second phase, the solution distribution scheme can take advantage of this solution representation. For instances, if we select fingerprinting constraints to a SAT instance such that 20 "don't-care" variables $x_0, x_1, \cdots, x_{19}$ are enforced, then every solution will be uniquely determined by a sequence of 20 bits, where all zero means the solution of $x_0 = x_1 = \cdots = x_{19} = 0$ and so on. Now for each IP buyer, we can encrypt the given signature file, hash it to a 20-bit integer, create a new solution based on this integer and assign it to the buyer.

We have made two basic fingerprinting assumptions: error-tolerance assumption and marking assumption. The first aims to guarantee a large solution space and the second enables collusion-free. Another fundamental question for fingerprinting is that *does the problem always have a large solution space and what happens if the solution space is very limited*? Since each user will receive a unique copy, we have to construct solution space large enough to accommodate all users or we are in trouble of releasing copies.

Many hard problems have a sufficient number of solutions in nature. For instance, in the GC problem, isolated nodes can be marked by any colors, and two connected nodes that have the same set of neighbors except themselves can exchange their colors. As another example, in the satisfiability (SAT) problem, flipping over the value of a don't-care variable in a satisfying assignment will give a different solution. For optimization problems, like GC, we can always get large solution space at the cost of solution quality degeneration. For decision problems like SAT, we cannot do much, however, fortunately, the solution space is usually huge except a few really hard instances[29]. Given that the solution space is large, to find k solutions is in general at least as hard as solving the original problem. Moreover, once we have the solution space, we have to maintain a one-to-one mapping from the solution to the user who receives this copy.

In our approach for solving the solution generation problem, we are not attempting to find the whole solution space. Instead, we add a set of extra constraints to the initial problem such that we can easily create (many) new solutions from one solution to the modified problem. In fact, we find a subspace of the solution space, where a base of this subspace can be built from this set of extra constraints, and the solution to the modified problem is a seed.

Once we have a set of solutions generated from a given base, where each solution can be uniquely expressed as a combination of the base. We can map each user's signature to a set of coefficients and assign him/her the corresponding copy of solution. Hence we only need to keep the base and the information for each user.

With a released solution, the user may gain some information about the problem. For example, if the user has a graph colored by 69 colors, then he knows the graph is 69-colorable and a satisfying assignment of a SAT problem tells the user that the original SAT is satisfiable. Since the solutions we created now are not random any more, users may collect different copies, detect their difference and produce new copies differ from their originals. The fingerprinting techniques should be designed to prevent this or make it hard, and allow the owner to be able to trace at least one of the dishonest users with a convincing probability from a forged copy.

4.3 Solution Creation Techniques

We present four techniques to generate solutions for the GC problem: (1) duplicating a selected set of vertices; (2) modifying small cliques; (3) adding edges between unconnected vertices; and (4) post-processing on one solution.

Vertex duplication

Given one coloring scheme to a graph, if we know that one vertex can also be colored by another alternative color, then immediately we can have one more solution to the same GC problem. Furthermore, on knowing k vertices each has an second valid color, we are able to create 2^k different solutions with almost no cost. And these k vertices and their associate colors will serve as the base for the solution space we have.

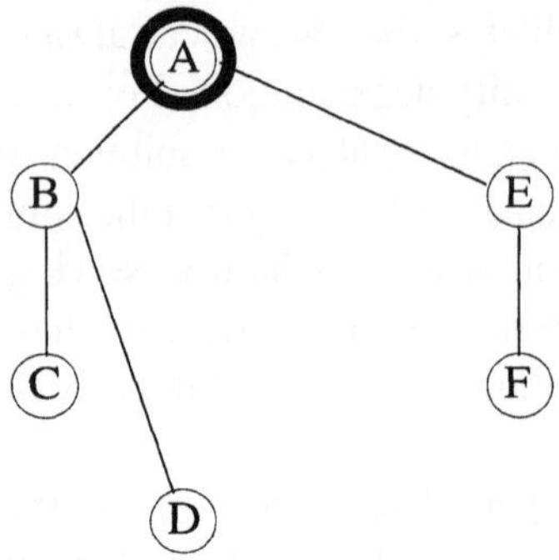

(a) Original graph.

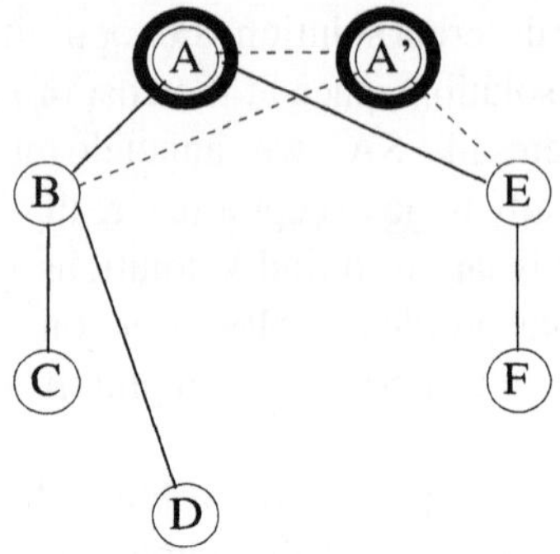

(b) New graph.

Figure 4.7. Duplicating vertex A to generate various solutions.

Figures 4.7 and 4.8 show this technique and an implementation. The idea is to select a vertex, duplicate it by creating a new vertex and connecting it to all the neighbor's of the selected vertex. Now the selected vertex can be labeled by either its color or the color of its duplication without violating the rules for GC. To guarantee these two vertices receive different colors, we add an edge in between. In Figure 4.7(b), vertices A and A' will be labeled by two different colors which can both be used to color A in the original graph 4.7(a).

Input: a graph $G(V, E)$, and an integer n. *Output:* a new graph G', such that n different solutions to G can be generated from one solution to G'.
Algorithm: copy $G(V, E)$ to $G'(V', E')$; copy V to V''; repeat ($\lceil \log_2 n \rceil$ times) { select a vertex v from V''; delete v from V''; add a new vertex v' to V'; add a new edge (v, v') to E'; for all (u such that edge $(u, v) \in E$) add an edge (u, v') to E'; } report graph $G'(V', E')$;

Figure 4.8. Pseudo code for vertex duplication.

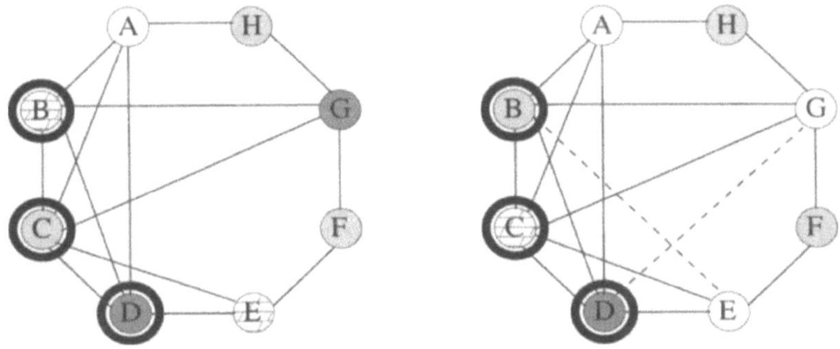

(a) Original graph. (b) New graph.

Figure 4.9. Manipulating small clique (triangle BCD).

Clique manipulation

In any valid color scheme, vertices from one clique will receive different colors, however, the solution may become invalid if they switch their colors. For example, consider the triangle BCD in Figure 4.9(a), once the other five vertices' colors are fixed as shown, it is easy to see this is the only solution.

We can add extra constraints to this triangle, as shown in Figure 4.9 (b), and now the three colors for vertices B, C, and D can be assigned arbitrarily. In general, if we choose a clique of size k, and for each vertex, we connect its neighbors to all other vertices in the clique, then based on one solution to the resulting graph, we get $k!$ solutions to the original GC problem by assigning each of the k different colors to one of the vertices in the clique.

Several cliques can be selected and they combine together forming a base for the solution space.

Bridge construction

There is no constraint for two vertices that do not have an edge connecting them. In [132], a watermarking technique is proposed where a message is embedded into the graph by adding edges between selected pairs of vertices, and the authorship can be claimed by showing the probability that every pair of vertices receiving different colors, which is not necessarily true in the original graph.

We can exploit the same idea here by selecting a pair of unconnected vertices, connecting one to all the neighbors of the other as well as these two vertices themselves. In Figure 4.10(b), vertices B and E are selected, and when we color the new graph, B and E will have different colors, say *red* and *green* . Now we can build 4 solutions where B and E are colored as *(red, red)*, *(red, green)*, *(green, red)* or *(green, green)*.

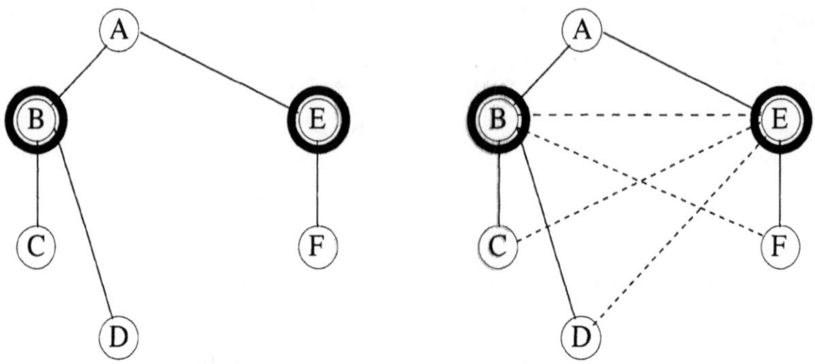

(a) Original graph. (b) New graph.

Figure 4.10. Constructing bridge between vertices B and E to generate various solutions.

It is worth mentioning here that this method is not restricted to a pair of unconnected vertices. We can select k unconnected vertices (an independent set of size k), create a complete graph over these vertices and connect each n ode to the neighbors of the others. Obviously, in this way, k^k different solutions can be derived from a single solution.

By constructing bridges, we can make the attacker's job very hard. In Figure 4.10, if two users detect that vertex B is marked by red and $green$ respectively in their solutions, and provided they know our fingerprinting technique, all the conclusion they may draw is that a bridge has been built between B and a vertex colored by either red or $green$. They have to search through a relatively large space and it will become even worse for them if we are selecting k unconnected vertices.

A hybrid of bridge construction and clique manipulation is practical with additional post-processing. We can choose k vertices (not necessarily unconnected), create a clique of size k and apply the clique manipulation technique. Now since the k selected vertices do not belong to an independent set, an arbitrary combination of their colors may not be valid in the original graph. A trivial procedure has to be conducted before releasing any solution which tests the validity of a given combination.

4.3.1 Solution post-processing

The last technique we discuss here requires post processing on a given solution.

Suppose we have colored graph $G(V, E)$ by k colors, denote V_i the subset of V that are colored by the ith color. So $V = \cup_{i=1}^{k} V_i$, and $V_i \cap V_j = \phi$ for all $i \neq j$. Now we select l colors and let $V' = \cup_{j=1}^{l} V_{i_j}$. Consider the subgraph $G[V']$ of G that is induced by V', we know this graph is l-colorable. In general, its size is relatively small and we can exhaustively find all the l-color solutions to it. Similarly we may construct another induced subgraph $G[V'']$ such that $V' \cap V'' = \phi$ and recolor it exhaustively. If we find n_1 and n_2 solutions for $G[V']$ and $G[V'']$ respectively, by applying the multiplication principle, we can create $n_1 \times n_2$ solutions to the original graph $G(V, E)$.

Comparison of the techniques:

One common characteristic for the first three techniques is that they belong to the category of *pre-processing*, where we modify the graph before it is colored by any GC solver (as a "blackbox"). Once a solution to the modified GC instance is returned, many different solutions to the original GC problem can be generated from this "seed" solution easily. The number of solutions can be controlled by tuning the parameters (see Table 1). But if we constrain the original graph too much, we may have some overhead, i.e., using extra colors for the modified graph comparing to that for the initial graph.

Technique and parameters	Number of solutions from one seed
Vertex duplication: duplicate k vertices	2^k
Clique manipulation: manipulate k cliques of size $n_1, n_2, \ldots, n_k$	$n_1! \times n_2! \times \ldots \times n_k!$
Bridge construction: select k sets of unconnected vertices of size $n_1, \ldots, n_k$	$n_1^{n_1} \times n_2^{n_2} \times \ldots \times n_k^{n_k}$
Solution post-processing: Partition the vertices into k subsets by their colors, in each subset, we find $n_1, n_2, \ldots, n_k$ solutions	$n_1 \times n_2 \times \ldots \times n_k$

Table 4.8. Summary of the four fingerprinting techniques.

In contrast, when we apply "solution post-processing" method, the GC solver will solve exactly the initial GC instance and it will provide us the best solution it can find. And in the post process, we always use the same amount of colors, therefore, there is guaranteed no overhead. However, it is not so easy to create many solutions as we do by the first three techniques, and the number of solutions are not controllable. In our experience, the better is the solver, the less space left for post-processing. For example, in a 85-color solution for a random graph of 1000 nodes, 66 colors are used for maximal independent sets.

We summarize these techniques in the following table, for each technique, we list its parameters and the size of the solution space. The base of the solution space can be easily built from the parameters, the overhead will be discussed later by experimental results.

4.4 Solution Distribution Schemes

As discussed before, the distributor wishes to give each user a uniquely fingerprinted copy. However, this is impractical for mass produced products like electronic books, software or CD-ROMs. One scheme[20] is to divide the data that a user received into two parts: the *public data* which is common to all users, and the *private data* which is unique to a particular user. Typically, the private part is small but should be able to provide enough information for the distributor to trace the user.

On the other hand, unlike human fingerprinting, the embedded digital fingerprints may be changed while the object is kept useful or functional correct. Two or more users may easily detect the difference between their copies, and come up with another copy without their fingerprints. In[20], for naive redistribution where a user redistributes his copy of the object without altering it, a *c-secure* code is constructed that can trace at least one of the guilty users from a coalition with size up to c users. For other cases, they construct *c-secure codes with* ε*-error* which allows an innocent user comes under suspicion with probability ε but requires a code length polynomial to $\log \frac{1}{\varepsilon}$ and $\log n$ (n is the number of potential users).

To avoid computing the problem many times, we create various solutions from one "seed" solution, therefore, similarities can be expected and it may be much easier for pirates to figure out these similarities and forge new valid solutions without their own fingerprints if the solutions are distributed improperly. For example, if we use the vertices duplication method with k vertices, in the seed solution, each of these k vertices will have a primary color and a secondary one. We are able to generate 2^k solutions where the only difference is the colors assigned to these k vertices. Suppose user A receives a copy of all the primary colors, and user B has one with all the secondary colors. Then if users A and B compare their copies, they can discover all the 2^k solutions.

We can discourage this with the aids of carefully designed distribution schemes. Although we cannot force users from redistribution, we can have the copies released in such a way that from a forged copy, we are able to catch at least one user from the coalition. The protocols in [18, 20] are applicable in this case. The basic idea is to select a subset of the solution space generated by the "seed solution" and release only solutions from this subset instead of the entire solution space. This subset should satisfy the following:

- Any combination of solutions cannot create a new solution in this subset, i.e., the innocent user will be protected.
- From any solution created by a combination of solutions from this subset, at lease one of the original solution can be traced. In another word, from an illegal copy, at least one of the guilty users will be caught.

Notice the domino effect of the GC problem (and many other hard optimization problems as well): changing the colors of a few vertices may render the entire solution. This phenomena does not exist in the contexts of fingerprints for classical objects, and our new techniques utilize it to discourage piracy. For example, if we use *clique manipulation* or *bridge construction* techniques, (or a hybrid of these two), it is still possible to find part or all the vertices that have been selected. However, the pirates will have difficult time to find the matching that tells them which clique it belongs to and/or which vertices are connected to it by bridges. And it is unlikely for the users to create new solutions, which are significantly different from the originals, from the copies generated by *solution post-processing*.

4.5 Experimental Results

We implement the proposed fingerprinting techniques in Section 4 on two types of graphs[175]. The first is standard random graphs with given number of vertices and edges. The other type of graphs is generated from the register allocation problem of variables in real codes. Table 2 shows the parameters for these graphs.

Fingerprinting random graphs

For the random graph DSJC1000.5.col.b, we color it on a Sun ULTRA-5 workstation and get five different solutions. Then we apply the proposed fingerprinting techniques on the original graph and color the resulting graphs again to get 5 solutions. The average and the best number of colors for each test are reported in Table 3. The last column shows the number of solutions can be derived from each single solution, recall that these solutions are guaranteed different. The run-time for coloring the original graph is about 16 hours, and those for the fingerprinted graphs are 14 $\sim$ 19 hours on the same system.

Though the run-time overhead can be ignored, the degradation of solution cannot. Graph DSJC1000.5.col.b has similar local structure everywhere by its nature. No matter which fingerprinting technique we use, we will make some part over-constrained and this causes the extra-color overhead.

Fingerprinting reallife benchmark graphs

To show the effectiveness of our proposed techniques, we fingerprint the reallife benchmark graphs in three different ways, which all promising different

Instance	Vertices	Edges	Optimal Coloring
fpsol2.i.1.col	496	11654	65
fpsol2.i.2.col	451	8691	30
fpsol2.i.3.col	425	8688	30
inithx.i.1.col	864	18707	54
inithx.i.2.col	645	13979	31
inithx.i.3.col	621	13969	31
mulsol.i.1.col	197	3925	49
mulsol.i.2.col	188	3885	31
mulsol.i.3.col	184	3916	31
mulsol.i.4.col	185	3946	31
mulsol.i.5.col	186	3973	31
zeroin.i.1.col	211	4100	49
zeroin.i.2.col	211	3541	30
zeroin.i.3.col	206	3540	30
DSJC1000.5.col.b	1000	249826	N/A

Table 4.9. Characteristics of benchmark graphs from real life.

Tests	Average	Best	Number of Solutions
Original Graph	85.8	85	1
Duplicate 10 Vertices	86.8	86	$2^{10} = 1024$
Duplicate 25 Vertices	87.8	87	$2^{25} \approx 3.355 \times 10^7$
Duplicate 50 Vertices	89.8	89	$2^{50} \approx 1.126 \times 10^{15}$
Create 5 Bridges	87.2	87	$4^5 = 1024$
Create 12 Bridges	89.2	89	$4^{12} \approx 1.678 \times 10^7$
Manipulate 4 Triangles	89.4	89	$6^4 = 1296$
Manipulate 10 Triangles	104	101	$6^{10} \approx 6.047 \times 10^7$

Table 4.10. Coloring the fingerprinted graph DSJC1000.5.col.b.

solutions to the order of 10^7 (10^{15} for *test2* and *test4*). Both original graphs and fingerprinted graphs can be colored in a few seconds on the same Sun ULTRA-5 workstation. The run-time overhead is negligible.

Table 3 reports the details on coloring the fingerprinted graphs. The first two columns are the instances and their optimal coloring. The next six columns are:

test1: select 25 vertices randomly and duplicate them. , capable of generating $2^{25} \approx 3.35 \times 10^7$ solutions.

test2: select 50 vertices randomly and duplicate them. , capable of generating $2^{50} \approx 1.12 \times 10^{15}$ solutions.

test3: repeat test1 with 25 carefully selected vertices. , capable of generating $2^{25} \approx 3.35 \times 10^7$ solutions.

test4: repeat test2 with 50 carefully selected vertices. , capable of generating $2^{50} \approx 1.12 \times 10^{15}$ solutions.

test5: apply bridge construction on 12 random pair of unconnected vertices. , capable of generating $4^{12} \approx 1.67 \times 10^7$ solutions.

test6: manipulate 10 random triangles. , capable of generating $6^{10} \approx 6.04 \times 10^7$ solutions.

Instance	opt.	test1	test2	test3	test4	test5	test6
# of solutions	1	2^{25} 3.3e07	2^{50} 1.1e15	2^{25} 3.3e07	2^{50} 1.1e15	4^{12} 1.6e07	6^{10} 6.0e07
fpsol2.i.1.col	65	66	66	*65*	*65*	*65*	78
fpsol2.i.2.col	30	31	34	*30*	*30*	*31*	42
fpsol2.i.3.col	30	31	32	*30*	*30*	*31*	44
inithx.i.1.col	54	54	56	*54*	*54*	*54*	60
inithx.i.2.col	31	33	33	*31*	*31*	32	41
inithx.i.3.col	31	33	34	*31*	*31*	32	40
mulsol.i.1.col	49	54	58	*49*	*49*	57	57
mulsol.i.2.col	31	37	38	*31*	*31*	36	36
mulsol.i.3.col	31	37	40	*31*	*31*	37	38
mulsol.i.4.col	31	37	39	*31*	*31*	35	40
mulsol.i.5.col	31	38	40	*31*	*31*	36	36
zeroin.i.1.col	49	53	56	*49*	51	53	60
zeroin.i.2.col	30	32	33	*30*	*30*	32	42
zeroin.i.3.col	30	32	32	*30*	*30*	33	43

Table 4.11. Coloring the fingerprinted real-life benchmark graphs.

In *test1* and *test2*, the overhead is significant, the reason is that we pick the vertices completely randomly. If we choose one from a clique of size k, a new clique of size $k+1$ will be created by duplicating a new vertex which makes the graph over-constrained. On the other hand, selecting isolated vertices only produce trivial solutions. Based on these observations, in *test3* and *test 4*, we avoid isolated vertices and those from large cliques. In all instances but one (*zeroin.i.1.col* with 50 vertices duplicated) there is no extra-color overhead.

The bridge construction method works fine for the *fpsol2* and *inithx* type of graphs, but bring unacceptable overhead to the other two. This is because that the *mulsol* and *zeroin* graphs are relatively small, consequently their solution spaces are small and to have the same amount of solutions, extra colors have to be introduced.

The clique manipulation technique is subtle than the previous ones, but it introduces overhead. When we select small cliques, most likely we will choose one from a large clique and possibly make the clique larger and the graph more difficult to be colored. For example, there are 5 triangles in Figure 4.11, one is the triangle on the right, the other four are from the clique of size 4. When we choose a triangle, with 80% we will pick one from the clique of size 4.

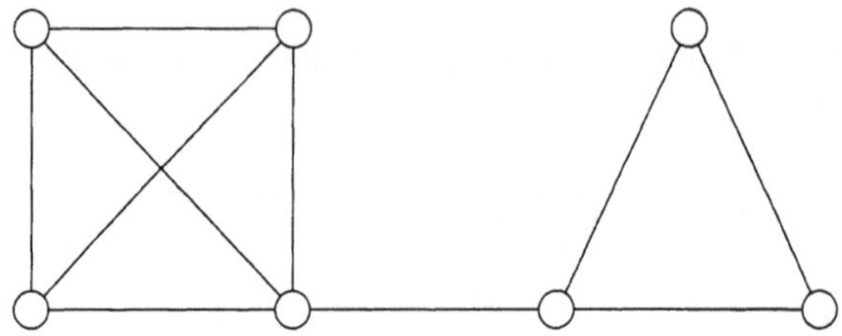

Figure 4.11. Choosing a triangle from a graph.

For real-life graphs, the local structure of the graph is different from place to place. More specifically, the constraints are not the same. We can exploit this unbalance and select (according to the owner's information if we want to watermark the solution as well) less-constrained part to apply the fingerprinting techniques. The above results show the effectiveness of this approach.

5. Summary

In this chapter, we discuss another part of the intellectual property protection, namely how to protect the right of legal IP buyers. In particular, we provide the symmetric fingerprinting techniques such that both IP provider's illegal distribution and IP buyers' collusion is discouraged. A fingerprinted IP will not directly prevent misuse of the IP, but will allow the IP provider to detect the source of the redistributed IP and therefore trace the traitor.

Fingerprinting-based IP has major advantages over watermarking-based intellectual property protection because it provides protection to both the buyer and seller. The key problem related to the use of fingerprinting for intellectual property protection is the tradeoff between collusion resiliency and runtime. Previous fingerprinting IP protection technique is applicable only to a very restricted set of problems[97].

We have introduced two generic fingerprinting technique for IP protection of solutions to optimization/decision problems and, therefore, of hardware and software intellectual property. By judiciously exploiting partial solution reuse and the incremental application of iterative optimizers, our first set of fingerprinting-based IP protection techniques for partitioning, graph coloring, satisfiability and placement, simultaneously provide high collusion resiliency and low runtime.

The second method enables fingerprinting at all level of design process, is applicable to an arbitrary optimization step, and produces numbers of distinct solutions with high quality. The key idea is to superimpose additional constraints on the problem formulation so to guarantee that the final solution can be in a straightforward way translated into k different high quality solutions. We have implemented this on the NP-complete GC problem and tested on a number of standard benchmarks. Fingerprinting random graphs introduces overhead, while for graphs generated from real-life register allocation problems, we have successfully created millions of distinct optimal solutions with no run-time overhead.

Notes

1 For example, suppose we have an object consisting of n real data values $v_1, v_2, \ldots, v_n$, and each value v_j has an associated delta value d_j, such that any number of $v_j - d_j, v_j$, and $v_j + d_j$ is acceptable for use by all users. Then immediately we can construct 3^n valid objects from one single set of n values.

2 For example, the Fiduccia-Mattheyses algorithm starts with a possibly random solution and changes the solution by a sequence of moves which are organized as *passes*. A move changes the assignment of a vertex from its current partition to another partition. At the beginning of a pass, all vertices are free to move (i.e., they are *unlocked*), and each possible move is labeled with the immediate change in total cost it would cause; this is called the *gain* of the move (positive gains reduce solution cost, while negative gains increase it). Iteratively, a move with highest gain is selected and executed, and the moving vertex is *locked*, i.e., is not allowed to move again during that pass. Since moving a vertex can change gains of adjacent vertices, after a move is executed all affected gains are updated. Selection and execution of a best-gain move, followed by gain update, are repeated until every vertex is locked. Then, the best solution seen during the pass is adopted as the starting solution of the next pass. The algorithm terminates when a pass fails to improve solution quality.

3 For some fingerprinting protocols, this can be useful for authentication. In the partitioning and standard-cell placement fingerprinting approaches below, which use weights rather than constraints, authentication will entail confirming that the solution IP is a local minimum with respect to a particular weighting (i.e., fingerprinted version) of the instance.

4 We use only one start since our CLIP FM implementation is deterministic; multiple starts from S_0 will yield the same local minimum.

5 We also mention that it is not required to color the fingerprinted graph G by the same GC algorithm $\mathcal{A}$ in step 4. This could pull the new solution further away from the initial solution S_0.

6 While some SAT solvers give only the truth variables and assume the rest are all false, other solvers do give don't care value to variables. If k variables are assigned don't cares in a solution, essentially this solution is equivalently to 2^k distinct solutions.

7 Recall that for a function $F(x_1, \cdots, x_n)$, its cofactor with respect to variable x_i is a function over $n-1$ variables such that

$$F_{x_i}(x_1, \cdots, x_{i-1}, x_{i+1}, \cdots, x_n) = F(x_1, \cdots, x_{i-1}, \mathbf{1}, x_{i+1}, \cdots, x_n).$$

Similarly, we can define

$$F_{x_i'}(x_1, \cdots, x_{i-1}, x_{i+1}, \cdots, x_n) = F(x_1, \cdots, x_{i-1}, \mathbf{0}, x_{i+1}, \cdots, x_n)$$

.

8 Thus, most entries in the table are non-integer.

9 In fact, additional constraints can be added when constructing the fingerprinting instances such that all existing solutions fail to satisfy the new fingerprinting instance. For example, if we get a truth assignment for the SAT problem: $x_1 = 0, x_2 = 1, x_3 = 1, \cdots, x_n = 0$, then adding the clause $x_1 + x_2' + x_3' + \cdots + x_n$ guarantees a distinct solution. However, how to create such constraints remains as another challenge.

Chapter 5

COPY DETECTION MECHANISMS FOR IP AUTHENTICATION

Clearly the success of digital watermarks and fingerprints relies on the detectability and traceability of the copyright marks. In this chapter, we present three different copy detection techniques. In the first approach, we choose signatures selectively and develop fast comparison schemes to detect such signatures. The second is a forensic engineering technique that identifies the source of an IP from a pool of sources based on their strategically different behavior. The last one is an enhanced detection-driven watermarking-fingerprinting method where part of the copyright marks are made public for easy-detection and cryptographic techniques for data integrity are applied to keep the marks secure and robust.

1. Introduction

The emergence of reuse-based design paradigm has improved the business model of semiconductor and EDA by the marketing (selling, renting, metering usage, etc.) of intellectual properties such as cores and EDA tools. Due to the already hot arena of legal disputes in the industry, it is believed that the main negative consequence of web-exposure of IP will be a significant increase of copyright infringement[90]. In such cases, the concerns of the plaintiffs are frequently related to the violation of patent rights accompanied with misappropriation of implemented software or hardware libraries. however, proving copyright obstruction has been a major obstacle in pursuing legal action and reaching a fair and convincing verdict. Needless to say, related losses, court rulings, or settlements have impacted enormously the market capitalization of involved companies. In fact, among more than 200 lawsuits filed by the software publishers association (merged to SIIA on January 1, 1999), all but one were settled out of court after the evidence of piracy has been discovered.

The *constraint-based IP protection* method consists of two main steps: the embedding of digital copyright marks such as watermark and fingerprint, and the detection of such marks. The first one prevents the unauthorized use of IP and the second enables the trace of unauthorized use if it occurs. Techniques for prevention, which are analogous to "locks" on the IP, include encryption, legal infrastructure, and closed infrastructures for IP dissemination. Techniques for detection, on the other hand, are aimed at discovering illegal copies of IP after the "locks" have been broken. In the VLSI CAD realm, the constraint-based watermarking approach prevents misappropriation by indelibly embedding the owner's signature into an IP, so that if an illegal copy is found the true owner's right can be established. However, the utility of watermarking is mostly after an illegal copy of IP has been found. The *copy detection problem* addresses the question of how to find the illegal copy in the first place.

We informally define the copy detection problem as follows[82]:

Given a library of n registered pieces of IP, and a new unregistered piece of IP, determine if any portion of any registered IP is present in the unregistered IP.

This definition reflects the use model for, say, a foundry which runs a copy detection program on any incoming design at the level of GDSII Stream representation. Copy detection is clearly complementary to existing watermarking-based IP protection techniques; below, we will show that it can also be enhanced by watermarking techniques.

There is a large body of research related to copy detection in several fields. Research on copy detection and plagiarism started in the early 1970s mainly as a technique for preventing widespread programming assignment copying[142] and to help support software reuse[83]. Over time a number of increasingly sophisticated techniques have been developed for programming assignment copy detection[66, 123, 161]. Most recently, even fractal and neural network-based techniques have been proposed for this task[117, 152].

In the database community, techniques for text copy detection have been developed[25, 106, 150]. A key approach is to find "signatures" (e.g., by hashing) of syntactically meaningful fragments (e.g., words or paragraphs), then create "term-document" or other incidence matrices that capture the presence of fragments within documents or IPs. Such incidence matrices are captured for all elements of a library of registered IPs. Then, when presented with a new IP, the copy detection system chunks the IP into fragments, and looks for matches of signatures in its library.

In the broader area of information hiding, most of the reported literatures on detection focus on how to extract and recover the embedded-data with the secret key from the stego-data, even after it has been attacked[124]. There are only two existing approaches to make watermark publicly detectable: one is based on the so-called public-key watermarking[72] and the other relies on zero-knowledge protocols[44]. We will review both later in this chapter.

Another area of related work is in string matching, which has received a great deal of attention since the early 1970s; see [2] for an excellent review. Several exceptionally effective algorithms have been proposed for rapid string matching in text[22, 83, 92]. For example, *awk* is a popular and powerful programming language that greatly facilitates development of tailored pattern scanning and processing software[171]. Finally, a number of copy detection techniques have been developed in biotechnology[15] and image processing[56].

Copy detection for VLSI CAD has been mainly performed at the layout level, where there is a need to eliminate or reduce redundant computation during VLSI artwork analysis (design rule checking, layout-versus-schematic (LVS) and pattern-based parasitic extraction). Techniques include isometry-invariant pattern matching[34, 116] and fast subgraph isomorphism algorithms[119]. Somewhat related work addresses template matching at various levels of the design process, where a design is covered by smaller templates available in a given library[41, 87].

2. Pattern Matching Based Techniques

The pattern matching based technique has the following elements[82]:

- For the given application domain, we identify a common structural representation of solutions (IPs), as well as what constitutes an element of the solution structure. Examples of such elements might include vertices in a netlist hypergraph, placed locations of edges in a custom layout, macros in a hierarchical GDSII Stream description of layout, steps in a schedule, and so on.

- For a given element type, we identify a means of calculating locally context dependent signatures for such elements, i.e., signatures that are functions of only an extremely local neighborhood of the element.

- Optionally, to speed comparison of IPs, we identify rare and/or distinguishing elements of a registered IP (cf. "iceberg queries" in [52]), and/or a hierarchy of signature types that may lead to faster filtering of negative (no match) comparisons.

- We develop fast (ideally, linear in the sizes of the IPs) comparison methods to identify suspicious unregistered IPs, e.g., by rare combinations of rare signatures. Subsequently, more detailed examination of suspicious IPs can be performed.

We define the objectives and methods for copy detection of programs used in system-level synthesis. An IP consists of a number of high-level procedures linked in an arbitrary fashion (e.g., DCT, vector motion compensation in MPEG). We assume:

- the adversary extracts a procedure or an entire library from the IP (e.g., DCT), and embeds the extracted code into his/her design;
- the adversary relinks the extracted procedures in an arbitrary fashion but without significant modification of the actual specification within each of the procedures; and
- the adversary may inline a procedure in the newly created specification or conduct peephole (local) perturbations.

We adopt this set of assumptions because of common risks involved in code obfuscation[38] and requirements for hardware-software maintenance (e.g. patches, incremental synthesis). The goal of the copy detection algorithm is to detect all procedures that have been copied from the original software. To perform this task, we have developed a copy detection mechanism operating at both the instruction selection level and the register assignment level; only the former is described here.

2.1 Copy Detection in High-Level Synthesis

We state the problem of copy detection for high-level synthesis as follows:

Given a set P of registered instruction sequences (procedures) of arbitrary lengths, and a suspected (i.e., suspicious) instruction sequence S, find the subset $P_0 \subseteq P$ consisting of all instruction sequences (procedures) $P_i \in P$ that occur in S (i.e., P_0 is a maximal subset such that $\forall P = i \in P_0, P_i \in S$).

To address this problem, we have developed an algorithm that uses probabilistic bounded search to identify copies. The algorithm is described in the pseudocode of Figure 5.1. we define a set of symbols A, the alphabet, which corresponds to the machine instruction set. Let p_i be the frequency of occurrence of symbol $a_i \in A$ in a given set P of code sequences. The algorithm initially determines the value of p_i for all a_i. Then, a subset $B \subseteq A$ of symbols from the alphabet is selected such that for each symbol $b_i \in B$ the probability of its occurrence is greater than zero and smaller than a predetermined constant $\epsilon(0 < p_i < epsilon)$, where ϵ is the bound for the probabilistic search.

For each procedure $P_i \in P$, the algorithm identifies the locations of all symbols from B. We consider "signatures" based on K-tuples of symbols from B. In particular, we find all K-tuples for which d, the maximum distance between any two elements of the K-tuple, is less than a prescribed value θ. The algorithm then creates a pattern *pat* for each such K-tuple. Due to the possibility of basic block reordering, the distance between two symbols is computed according to the distance in the dynamic execution. In addition, due to possible instruction

```
Find probability p_i of occurrence of each symbol a_i from
  alphabet A in a given set P of code sequences.
Select a subset B of symbols from alphabet A such that
  ∀a_i = inB, 0 < p_i < ε.
PoolPatterns = φ.
For each P_i ∈ P
  Identify locations of all symbols from B.
  Create a pattern pat for each K-tuple of symbols from B,
    with max distance d among elements of the K-tuple (d <= θ).
  Add pat(symbols, distanceMatrix) to PoolPatterns.
EndFor   Select the least frequent M from PoolPatterns.
Find a subset C ⊆ B of symbols that covers all patterns from
  PoolPatterns_M and has the smallest sum of occurrence
  probabilities.
-----------------------------------------------------------------
For each element c ∈ C
  if c ∈ S
    For each pattern pat∈ PoolPatterns_M that contains c
      if corresponding symbols are at locations
        specified by c.distanceMatrix signal pat is found.
        The locations of symbols are verified in increasing order
          of the symbol's probability of occurrence.
      if pat is found
        Perform pattern matching to verify copy in [68] and [83].
```

Figure 5.1. Pseudo-code for software copy detection at the instruction selection level (preprocessing and detection).

reordering, symbols are not searched at exact distances, but within a neighborhood (of cardinality N) of the exact location[1]. Parameters ϵ, K, N, and θ are selected such that all procedures from P contain at least one pattern. The probability that a specific pattern appears in a code sequence is:

$$Prob\{pat \subset S\} = \sum_{i=1}^{K} \prod_{j=1}^{i} (1 - p_{c_j \in pat})^N)$$

All identified patterns are stored in a pool of patterns, *PoolPatterns*. Each pattern is represented using its symbols and the matrix that specifies the distances between symbols. To reduce the sample of IP code selected for comparison, the algorithm selects a setofM least frequent patterns from *PoolPatterns* that cover all procedures from P; this is called the *constrained PoolPatterns set.* The algorithm also identifies a subset $C \subseteq B$ of symbols that cover all patterns from the constrained *PoolPatterns* and has the smallest sum of symbol occurrence probabilities.

Finally, the suspected sequence of instructions is sequentially parsed for symbols from C. If a symbol $c \in C$ is found, all patterns that contain c are matched using their distance matrices for occurrence of the remaining symbols. The remaining symbols are searched in the order of their occurrence probabilities. If a specific pattern is identified in S, the algorithm performs an exact pattern matching of all procedures that contain *pat* and S to verify the copy detection signal[83], or else performs non-exact pattern matching using the *diff* utility program[68].

2.2 Copy Detection in Gate-Level Netlist Place-and-Rout

In the automated place-and-route domain, we seek to protect a gate-level cell netlist that may contain embedded placement information. Such a design artifact typically arrives in Cadence Design Systems, Inc. LEF/DEF interchange format; we parse this to yield a netlist hypergraph with pin direction information. The fundamental test for netlist copying is *isomorphism checking*, i.e., finding subhypergraphs of one (unregistered) netlist in another (registered) netlist. Isomorphism checking is essentially near-linear time for rigid graphs, i.e., graphs without automorphisms – and this includes almost all graphs (cf., e.g., [119] in the VLSI CAD literature). Nevertheless, we must still filter calls to isomorphism checkers, because there are so many subhypergraphs that are potentially subject to copying. Filtering depends on (a hierarchy of) comparisons that span a continuum between "coarse" and "detailed", and is what enables practically useful methods. For example, checking whether two chips' netlists have the same number of cells, same number of macro types, same sorted cell degree sequences, same number of connected components, etc. are all coarse but potentially effective comparisons; checking isomorphism is a detailed comparison.

The filtering approach[82] is based on finding a "signature" for each individual cell (i.e., vertex in the netlist hypergraph) using a simple encoding of the cell's neighborhood. Specifically, we record for cell c_i the sequence of values[2]:

- $|N_{i,1}|$ = the cardinality of the set of distinct nets incident to c_i,
- $|C_{i,1}|$ = the cardinality of the set of distinct cells on the nets in $\{N_{i,1}\}$, and
- $|N_{i,2}|$ = the cardinality of the set of distinct nets incident to the cells in $\{C_{i,1}\}$, etc.

Several practical considerations arise. (1) Because the diameter of a netlist hypergraph is not large, and because we would like such signatures to identify specific cells even in a small fragment of the original netlist, we record only the first k elements of this sequence (in our experiments below, we use $k = 6$). On the other hand, to increase the likelihood that such sequences can uniquely determine a match, we actually compute such sequences in several variants of the hypergraph, corresponding to deleting hyperedges whose degree exceeds some threshold d. (In the experiments below, we generate three sequences for each cell, corresponding to $d = 4, 7, 10$.) We also break each entry of the sequence into subentries according to pin direction (in, out, in-out). Thus, there are $6 \times 3 \times 3 = 54$ numbers in each cell's sequence. (2) Finding one match of all 54 numbers in a sequence is much rarer than, say, three different matches of 18 numbers. To capture this, we give geometrically more *credit* for a longer match, e.g., *credit*= $2^{\lfloor (x-1)/9 \rfloor}$, where x is the number of positions in which two sequences' entries match. (3) Finally, because we do not wish to spend CPU time comparing all cell sequences from the unregistered IP against all cell sequences from the registered IP, we lexicographically order the entries of the 54-number sequences with all entries due to $i = 1$ before entries due to $i = 2$, etc. Furthermore, we adopt the convention that the number of positions in which two sequences match is simply given by the length of the longest common prefix of both sequences. In this way, finding the *best* matches for *all* sequences of the unregistered IP, within the list of sequences for the registered IP, is accomplished in linear time by pointer-walking in two sorted lists. Hence, we do not need to resort to use of "rare" signatures for complexity reduction.

We have also considered copy detection in polygon layouts that may have been exposed to migration and compaction tools during copying. We initially filter macros by signatures according to simple attributes (number of features per layer, size, etc.). A second filter (before actual isomorphism checking) uses vertex signatures in "conflict graphs" defined over features in the layout; in a conflict graph, the number of vertices equals the number of layout features, and there is an edge between vertices if corresponding features are within distance d of each other (varying d induces a family of such graphs). When d is significantly larger than the minimum feature size/spacing, then slight changes in layout will not affect the conflict graph.

2.3 Experimental Results

We have performed a set of experiments to evaluate the effectiveness of the copy detection mechanism for behavioral specifications. We use the standard multimedia benchmark applications[101], Sun's UltraSparc instruction set and its instruction-set simulator SHADE. In the preprocessing step, for the set of applications shown in Table 5.1, we identify the distribution of occurrence of instructions as well as the required distance matrices for all established patterns (cf. *http://www.cs.ucla.edu/ leec/mediabench/applications.html* for the detailed histograms). Because the performance of the copy detection mechanism is by and large based on the statistical analysis of the IP code, the approach performs lengthy explorations in the pre-processing step with an objective to increase the performance of the algorithm (i.e., lower $Prob\{pat \subset S\}$). While the pre-processing step took, on the average, 46 hours for a single application, the actual detection process required in all experimental cases is less than 10 seconds. Table 5.1 shows the obtained results for the detection process. Column 1 shows the name of the application; Column 2 shows the size of the suspected code and the number of procedures; Column 3 shows the number of "original" procedures; Column 4 shows the cumulative probability of false alarm P_{FA}; and Column 5 shows the probability of detection was 100%. As presented in Table 5.1, the probability of a false alarm, accumulated for all considered patterns, quantifies the performance of the algorithm because it is proportional to the number of negative tests due to exact pattern matching.

Application	Suspected Code Size	IP Procedures	P_{FA}	CPU time for detection
JPEG encoding	391	24	$3.3 \cdot 10^{-7}$	1.8s
JPEG decoding	379	24	$1.8 \cdot 10^{-8}$	1.2s
PGP encryption	443	36	$2.1 \cdot 10^{-9}$	3.2s
MPEG decoding	114	17	$5.2 \cdot 10^{-11}$	2.9s
G.721 encoding	26	4	$9.1 \cdot 10^{-5}$	0.0s
GSM encoding	98	8	$8.1 \cdot 10^{-7}$	0.7s

Table 5.1. Effectiveness of the copy detection mechanism for behavioral specifications.

Weighted Matching Between Full Designs						
	A	B	C	D	E	F
A	100%	7.99%	2.90%	4.19%	2.84%	2.84%
B	2.89%	100%	1.27%	2.12%	1.26%	1.26%
C	0.77%	0.76%	100%	1.20%	0.72%	0.73%
D	2.69%	2.79%	0.42%	100%	0.33%	0.33%
E	0.25%	0.36$	0.25%	0.24%	100%	30.3%
F	0.16%	0.27%	0.16%	0.15%	28.7%	100%

Table 5.2. Matching percentage between two full designs, based on weighted sum of credits. The matching percentage between Cases E and F may be high because of potential reused IP between these designs.

We have also applied the copy detection procedure discussed above to compute cell sequences for 6 industry standard-cell designs in LEF/DEF format. The number of cells in the designs (Cases A - F) are respectively 3286, 12133, 12857, 20577, 57275 and 117617. Cases E and F are from the same design team and may contain common subdesigns. Table 5.2 shows the total matching credits when Case i is matched into Case j, i.e., the best match for each cell in Case i is found within Case j. Table 5.3 shows the total matching credits when a portion of Case i (a connected component of 500 cells, found by breadth first search from a randomly chosen cell) is matched into Case j. (Here, the results are averaged over three separate trials.) We express the total matching credit as a percentage of the maximum possible total credit. In our current use model, all registered IPs are checked against the unregistered IP. Hence, we are able to see which IPs have higher matching credits relative to the other IPs. Typically, matching percentages for non-copied IPs are in a fairly narrow range, while those of copied IPs are significantly higher. Note that in Tables 5.2 and 5.3, there was a big difference between matching of Case E and Case F and matching between any other case and Case F. Larger IPs will tend to afford better distinction between copied IPs and non-copied IPs, as seen by comparing the two Tables.

Weighted Matching of Partial vs. Full Designs						
Partial	Full Designs					
Designs	A	B	C	D	E	F
A	32.6%	8.24%	6.41%	7.21%	6.28%	6.28%
B	5.89%	14.6%	4.80%	6.46%	4.59%	4.59%
C	4.03%	4.03%	19.3%	4.40%	3.98%	3.98%
D	11.2%	12.7%	10.8%	23.6%	10.6%	10.6%
E	6.46%	6.50%	6.43%	6.41%	13.6%	10.1%
F	5.49%	5.63%	5.45%	5.44%	7.13%	15.8%

Table 5.3. Percentage of matching between partial design and full design with weighted sum of the credits. Each entry is an average over three experimental trials.

3. Forensic Engineering Techniques

3.1 Introduction

Forensic analysis is a key methodology in many scientific and art fields, such as anthropology, science, literature, and visual art. For example, forensics is most commonly used in DNA identification. Rudin et al. present the details on DNA profiling and forensic DNA analysis[140]. In literature Thisted and Efron used statistical analysis of Shakespeare's vocabulary throughout his works to predict if a new found poem came from Shakespeare's pen[160]. They provided a high confidence statistical argument for the positive conclusion by analyzing how many new words, words used once, twice, three times and so on would appear in the new Shakespeare's work.

Software copyright enforcement has attracted a great deal of attention among law professionals. McGahn gives a good survey on the state-of-the-art methods used in court for detection of software copyright infringement[110]. In the same journal paper, McGahn introduces a new analytical method, based on

Learned Hand's abstractions test, which allows courts to rely their decisions on well established and familiar principles of copyright law. Grover presents the details behind an example lawsuit case[67] where Engineering Dynamics Inc. is the plaintiff issuing a judgment of copyright infringement against Structural Software Inc., a competitor who copied many of the input and output formats of Engineering Dynamics Inc.

Forensic engineering has received little attention among the computer science and engineering research community. To the best knowledge of the authors, to date, forensic techniques have been explored for detection of authentic Java byte codes[10] and to perform identity or partial copy detection for digital libraries[25]. Recently, steganography and code obfuscation techniques have been endorsed as viable strategies for content and design protection. We have seen the constraint-based watermarking and fingerprinting methods in the previous chapters for the protection of VLSI design IPs. In the software domain, good survey of techniques for copyright protection of programs has been presented by Collberg and Thomborson[38, 40]. They have also developed a code obfuscation method which aims at hiding watermarks in program's data structures. Although steganography and obfuscation have demonstrated potential to protect software and hardware implementations, their applicability to algorithm protection is still an unsolved issue. In order to provide a foundation for associating algorithms with their creations, techniques aiming at detecting copyright infringement by giving quantitative and qualitative analysis of the algorithm-solution correspondence.

3.2 Forensic Engineering for the Detection of VLSI CAD Tools

3.2.1 Generic Approach

Forensic engineering aims at providing both qualitative and quantitative evidence of substantial similarity between the design original and its copy. The generic problem that a forensic engineering methodology tries to resolve can be formally defined as follows:

> *Give a solution S_P to a particular optimization problem instance P and a finite set of algorithms A applicable to P, the goal is to identify with a certain degree of confidence which algorithm A_i has been applied to P for obtaining solution S_P.*

An additional restriction is that the algorithms (their software or hardware implementations) have to be analyzed as black boxes. This requirement is based on two facts: (i) similar algorithms can have different executables and (ii) parties involved in the ruling are not eager to reveal their IP even in court. The global flow of the generic forensic engineering approach consists of three fully modular phases:

Statistics collection. Initially, each algorithm $A_i \in A$ is applied to a large number of isomorphic representations $P_j, j = 1, \cdots, N$ of the original problem instance P. Note that "isomorphism" indicates pseudo-random perturbation of the original problem instance P. Then, for each obtained solution $S^i_{P_j}, i = 1, \cdots, |A|, j = 1, \cdots, M$, an analysis program computes the values $w^{i,j}_k, k = 1, \cdots, L$ for a particular set of solution's properties $\pi_k, k = 1, \cdots, L$. The reason behind performing iterative optimizations of perturbed problem instances is to obtain a valid statistical model on certain properties of solutions generated by a particular algorithm.

Next, the collected statistical data $\{w^{i,j}_k\}$ is integrated into a separate histogram χ^i_k for each property π_k under the application of a particular algorithm A_i. Since the probability distribution function for χ^i_k is in general not known, using non-parametric statistical methods[48], each algorithm A_i is associated with probability $p_{\chi^i_k} = x$ that its solution results in property π_k being equal to x.

Algorithm clustering. In order to associate an algorithm $A_x \in A$ with the original solution S_P, the set of algorithms is clustered according to the properties of S_P. The value $w^{S_P}_k$ for each property π_k of S_P is then compared to the collected histograms (χ^i_k, χ^j_k) of each pair of considered algorithms A_i and A_j. Two algorithms A_i and A_j remain in the same cluster if the likelihood that their properties are not correlated it greater than some predetermined bound ϵ.

It is important to stress that a set of properties associated with algorithm A_i can be correlated with more than one cluster of algorithms. For instance, this can happen when an algorithm A_i is a blend of two different heuristics (A_j, A_k) and therefore its properties can be statistically similar to the properties of A_j and A_k. Obviously, in such cases exploration of different properties or more expensive and complex structural analysis of programs is the only solution.

Decision making. If the plaintiff's algorithm A_x is clustered jointly with the defendant's algorithm A_y and A_y is not clustered with any other algorithm from A, substantial similarity between the two algorithms is positively detected. The selection of properties plays an important role in the entire system. Two obvious candidates are the actual quality of solution and the run-time of the optimization program. Needless to say, such properties may be a decisive factor only in specific cases when copyright infringement has not occurred. Only detailed analysis of solution structures can give useful forensic insights.

We explain the detailed forensic approach for graph coloring and Boolean satisfiability problems next.

3.2.2 Statistics Collection for Graph Coloring Problem

Due to the importance of the graph coloring problem and its numerous applications, there exist a number of exact and heuristic algorithms. We select the following solvers as the pool of algorithms A for brevity and due to the limited accessibility to the source code: greedy, DSATUR, RLF-based MAXIS, backtrack DSATUR, iterated greedy, and tabu search described in [172].

The simplest constructive algorithm for graph coloring is the "sequential" coloring algorithm (SEQ). SEQ sequentially traverses and colors vertices with the lowest index not used by the already colored neighboring vertices. DSATUR [23] colors the next vertex with a color C selected depending on the number of neighbor vertices already connected to nodes colored with C (saturation degree) as shown in Figure 5.2. RLF [102] colors the vertices sequentially one color class at a time. Vertices colored with one color represent an independent subset (IS)of the graph. The algorithm tries to color with each color maximum number of vertices. Since the problem of finding the maximum IS is intractable, a heuristic is employed to select a vertex to join the current IS as the one with the largest number of neighbors already connected to that IS. An example how RLF colors graphs is also presented in Figure 5.2. Node 6 is randomly selected as the first node in the first IS. Two nodes (2,4)have maximum number of neighbors which are also neighbors to the current IS. The node with the maximum degree is chosen (4). Node 2 is the remaining vertex that can join the first IS. The second IS consists of randomly selected node 1 and the only remaining candidate to join the second IS, node 5. Finally, node 3 represents the last IS.

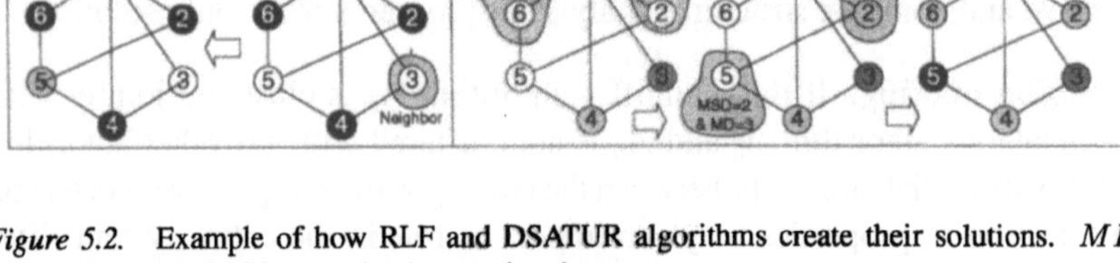

Figure 5.2. Example of how RLF and DSATUR algorithms create their solutions. MD - maximal degree; MSD - maximal saturation degree.

Iterative improvement techniques try to find better colorings through generating successive colorings by random moves. The most common search techniques are simulated annealing and tabu search[163, 55]. In our experiments, we will use XIS (RLF based), backtrack DSATUR, iterated greedy, and tabu search.

A successful forensic technique should be able to, given a colored graph, distinguish whether a particular algorithm has been used to obtain the solution. The key to the efficiency of the forensic method is the selection of properties used to quantify algorithm-solution correlation. We use a list of properties that aim at analyzing the structure of the solution:

π_1: Color class size. Histogram of IS cardinalities is used to filter greedy algorithms that focus on coloring graphs constructively (e.g. RLF-like algorithms). Such algorithms tend to create large initial independent sets at the beginning of their coloring process. To quantify this property, we take the cardinality of the largest IS normalized against the size of the average IS in the solution. Alternatively, as a slight generalization, in order to achieve statistical robustness, we use 10% of the largest sets instead of only the largest. Interestingly, on real-life applications the first metric is very effective, and on random graphs the second one is strong indicator of the used coloring algorithm.

π_2: Number of edges in large independent sets. This property is used to aid the accuracy of π_1 by excluding easy-to-find large independent sets from consideration in the analysis. We use $k\%$ of the largest sets and measure the percentage of edges leaving the IS.

π_3: Number of edges that can switch color classes. This criteria analyzes the quality of the coloring. Good (in a sense of being close to a local minima)coloring result will have fewer nodes that are able to switch color classes. It also characterizes the greediness of an algorithm because greedy algorithms commonly create at the end of their coloring process many color classes that can absorb large portion of the remaining graph. The percentage of nodes which can switch colors versus the number of nodes is used.

π_4: Color saturation in neighborhoods. This property assumes creation of a histogram that counts for each vertex the number of adjacent nodes colored with one color. Greedy algorithms and algorithms that tend to sequentially traverse and color vertices are more likely to have node neighborhoods dominated by fewer colors. We want to know the number of colors in which the neighbors of any node are colored. The Gini coefficient is used as well as the average value to quantify this property. The Gini coefficient is a measure of dispersion within a group of values, calculated as the average difference between every pair of values divided by two times the average of the sample. The larger the coefficient, the higher the degree of dispersion.

π_5: Sum of degrees of nodes included in the smallest color classes. This property aims at identifying algorithms that perform peephole optimizations, since they are not likely to create color classes with high-degree vertices.

π_6: Sum of degrees of nodes adjacent to the vertices included in the smallest color classes. The analysis goal of this property is similar to π_5 with the exception that it focuses on selecting algorithms that perform neighborhood look ahead techniques[88]. The values are normalized against the average value and both the average value and the Gini coefficients are used.

π_7: Percent of maximal independent subsets. This property can be highly effective in distinguishing algorithms that color graphs by iterative color class selection (RLF). Supplemented with property π_3, it aims at detecting fine nuances among similar RLF-like algorithms.

The itemized properties can be effective only on large instances where the standard deviation of histogram values is relatively small. Using standard statistical approaches, the function of standard deviation for each histogram can be used to determine the standard error in the reached conclusion.

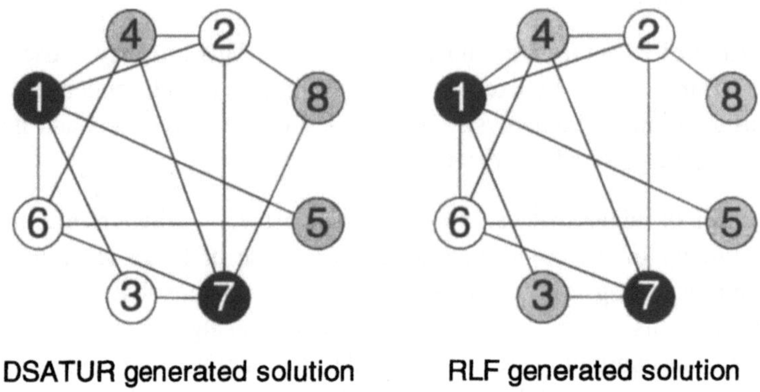

Figure 5.3. Example of two different graph coloring solutions obtained by two algorithms DSATUR and RLF. The index of each vertex specifies the order in which it is colored according to a particular algorithm.

Although instances with small cardinalities cannot be a target of forensic methods, we use a graph instance in Figure 5.3 to illustrate how two different graph coloring algorithms tend to have solutions characterized with different properties. The applied algorithms are DSATUR and RLF. Specified algorithms color the graph constructively in the order denoted in the figure. If property π_1 is considered, the solution created using SATUR has a histogram $\chi_{\pi_1}^{DSATUR} = \{1_2, 2_3, 0_4\}$, where histogram value x_y denotes x sets of color classes with cardinality y. Similarly, the solution created using RLF results $\chi_{\pi_1}^{RLF} = \{2_2, 0_2, 1_4\}$. Commonly, extreme values point to the optimization goal of the algorithm or characteristic structure property of its solutions. In this case, RLF has found a maximum independent set of cardinality $y = 4$, a consequence of algorithm's strategy to search in a greedy fashion for maximal ISs.

3.2.3 Statistics Collection for Boolean Satisfiability Problem

There are at least three broad classes of solution strategies for the SAT problem. The first class of techniques are based on probabilistic search[151, 144], the second are approximation techniques based on rounding the solution to a nonlinear program relaxation[62], and the third is a great variety of BDD-based techniques[26]. We select the following SAT algorithms to demonstrate our forensic engineering technique:

- GSAT. It identifies for each variable v the difference $DIFF$ between the number of clauses currently unsatisfied that would be satisfied if the truth value of v were reversed and the number of clauses currently satisfied ed that would become unsatisfied if the truth value of were flipped[145]. The algorithm pseudo-randomly flips assignments of variables with the greatest $DIFF$.
- WalkSAT. It selects with probability p a variable occurring in some unsatisfied clause and flips its truth assignment. Conversely, with probability $1 - p$, the algorithm performs a greedy heuristic such as GSAT[147].
- NTAB. It performs a local search to determine weights for the clauses, intuitively giving higher weights corresponds to clauses which are harder to satisfy. The clause weights are then used to preferentially branch on variables that occur more often in clauses with higher weights[46].
- Rel_SAT_rand. It represents an enhancement of GSAT with look-back techniques[11].

In order to correlate an SAT solution to its corresponding algorithm, we have explored the following properties of the solution structure:

π_1: Percentage of non-important variables. A variable v is *non-important* for a particular set of clauses C and satisfactory truth assignment $t(V)$ of all variables in V, if both assignments $t(v) = T(rue)$ and $t(v) = F(alse)$ result in satisfied C. For a given truth assignment t, we denote the subset of variables that can switch their assignment without impact on the Satisfiability of C as V_{NI}^t. In the remaining set of properties only functionally significant subset of variables $V_0 = V - V_{NI}^t$ is considered for further forensic analysis.

π_1: Clausal stability. Clausal stability is the percentage of variables that can switch their assignment such that $k\%$ of clauses in C are still satisfied. This property aims at identifying constructive greedy algorithms, since they assign values to variables such that as many as possible clauses are covered with each variable selection.

π_3: Ratio of true assigned variables vs. total number of variables in a clause. Although this property depends by and large on the structure of the problem, in general, it aims at qualifying the effectiveness of the algorithm. Large values commonly indicate usage of algorithms that try to optimize the coverage using each variable.

π_4: Ratio of coverage using positive and negative appearance of a variable. While property π_3 analyzes the solution from a perspective of a single clause, this property analyzes the solution from a perspective of each variable. Each variable v_i appears in p_i clauses as positively and n_i clauses as negatively inclined. The property quantifies the possibility that an algorithm assigns a truth value to $t(v_i) = p_i \geq n_i$.

π_5: The GSAT heuristic. For each variable v, the difference $DIFF = a - b$ is computed, where a is the number of clauses currently unsatisfied that would become satisfied if the truth value of v were reversed, and b is the number of clauses currently satisfied that would become unsatisfied if the truth value of v were flipped. This measure only applies to maximum SAT problems, where the problem is to find the maximum number of clauses which can be satisfied at once.

As in the case of graph coloring, the listed properties demonstrate significant statistical proof only for large problem instances. Instances should be large enough to result in low standard deviation of collected statistical data.

3.2.4 Algorithm Clustering and Decision Making

Once statistical data is collected, algorithms in the initial pool are partitioned into clusters. The goal of partitioning is to join strategically similar algorithms (e.g. with similar properties) into a single cluster. This procedure is presented formally using the pseudo-code in Figure 5.4.

The clustering process is initiated by setting the starting set of clusters to empty $C = \emptyset$. In order to associate an algorithm $A_x \in A$ with the original solution S_P, the set of algorithms is clustered according to the properties of S_P. For each property π_k of S_P we compute its feature quantifier $\pi_k(S_P) \rightarrow \omega_k^{S_P}$ and compare it to the collected pdfs of corresponding features χ_k^i of each considered algorithm $A_i \in A$. The clustering procedure is performed in the following way: two algorithms A_i, A_j remain in the same cluster, if the likelihood $z(A_i, A_j)$ that their properties are not correlated is greater than some predetermined bound $\epsilon \ll 1$.

$$z(A_i, A_j) = \prod_{k=1}^{|\pi|} \frac{2 \cdot \min(\Pr[\pi_k(A_i) \rightarrow \omega_k^i], \Pr[\pi_k(A_j) \rightarrow \omega_k^j])}{\Pr[\pi_k(A_i) \rightarrow \omega_k^i] + \Pr[\pi_k(A_j) \rightarrow \omega_k^j]}$$

The function that computes the mutual correlation of two algorithms takes into account the fact that two properties can be mutually dependent. Algorithm A_i is added to a cluster C_k if its correlation with all algorithms in C_k is greater than some predetermined bound $\epsilon \leq 1$. If A_i cannot be highly correlated with any algorithm from all existing clusters in C then a new cluster $C_{|C|+1}$ is created with A_i as its only member and added to C. If there exists a cluster C_k for which A_i is highly correlated with a subset C_k^H of algorithms within C_k, then C_k is partitioned into two new clusters $C_k^H \cup A_i$ and $C_k - C_k^H$. Finally, algorithm A_i is removed from the list of unprocessed algorithms A. These steps are iteratively repeated until all algorithms are processed.

```
Given A. C = ∅.
For each A_i ∈ A
  For each C_k ∈ C
    add = true; none = true
    For each A_j ∈ C_k
      If z(A_i, A_j) ≤ ε.
      Then add = false Else none = false
    End For
    If add Then merge A_i with C_k
    Else create new cluster C_{|C|+1} with
    A_i as its only element.
    If none Then create two new clusters
    C_k^H ∪ A_i and C_k − C_k^H where C_k^H ∈ C_k
    is a subset of algorithms highly correlated with A_i.
  End For
End For
```

Figure 5.4. Pseudo-code for the algorithm clustering procedure.

According to this procedure, an algorithm A_i can be correlated with two different algorithms A_j, A_k that are not mutually correlated (as presented in Figure 5.5). For instance, this situation can occur when an algorithm A_i is a blend of two different heuristics (A_j, A_k) and therefore its properties can be statistically similar to the properties of A_j, A_k. In such cases, exploration of different properties or more expensive and complex structural analysis of algorithm implementations is the only solution to detecting copyright infringement.

Obviously, according to this procedure, an algorithm A_i can be correlated with two different algorithms A_j, A_k that are not mutually correlated (as presented in Figure 6). For instance this situation can occur when an algorithm A_i is a blend of two different heuristics (A_i, A_k) and therefore its properties can

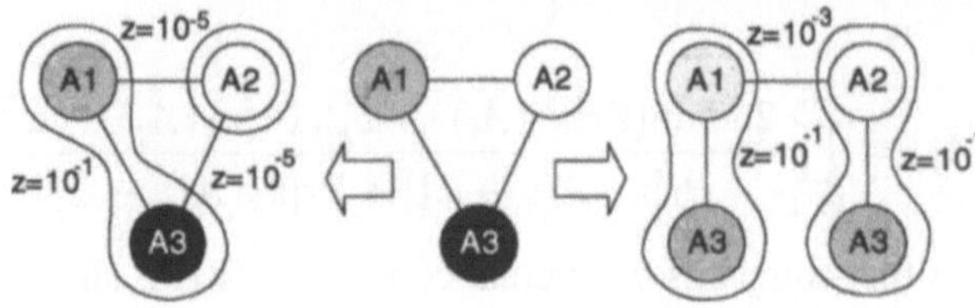

Figure 5.5. Two different examples of clustering three distinct algorithms. The first clustering (figure on the left) recognizes substantial similarity between algorithms A_1 and A_3 and substantial dissimilarity of A_2 with respect to A_1 and A_3. Accordingly, in the second clustering (figure on the right) the algorithm A_3 is recognized as similar to both algorithms A_1 and A_2, which were found to be dissimilar.

be statistically similar to the properties of A_j, A_k. In such cases, exploration of different properties or more expensive and complex structural analysis of algorithm implementations is the only solution to detecting copyright infringement. Once the algorithms are clustered, the decision making process is straightforward:

If plaintiff's algorithm A_x is clustered jointly with the defendant's algorithm A_y (e.g. its solution S_P)
and A_y is not clustered with any other algorithm from A which has been previously determined as strategically different,
then substantial similarity between the two algorithms is positively detected at a degree quantified using the parameter $z(A_x, A_y)$.

The court may adjoin to the experiment several slightly modified replicas of A_x as well as a number of strategically different algorithms from A_x in order to validate that the value of $z(A_x, A_y)$ points to the correct conclusion.

3.3 Experimental Results

In order to demonstrate the effectiveness of the proposed forensic methodologies, we have conducted a set of experiments on both abstract and real-life problem instances. In this section, we present the obtained results for a large number of graph coloring and SAT instances. The collected data is partially presented in Figure 5.6. It is important to stress, that for the sake of external similarity among algorithms, we have adjusted the run-times of all algorithms such that their solutions are of approximately equal quality.

We have focused our forensic exploration of graph coloring solutions on two sets of instances: random (1000 nodes and 0.5 edge existence probability) and register allocation graphs. The last five subfigures in Figure 5.6 depict the histograms of property value distribution for the following pairs of algorithms and properties: DSATUR with backtracking vs. maxis and π_3, DSATUR with

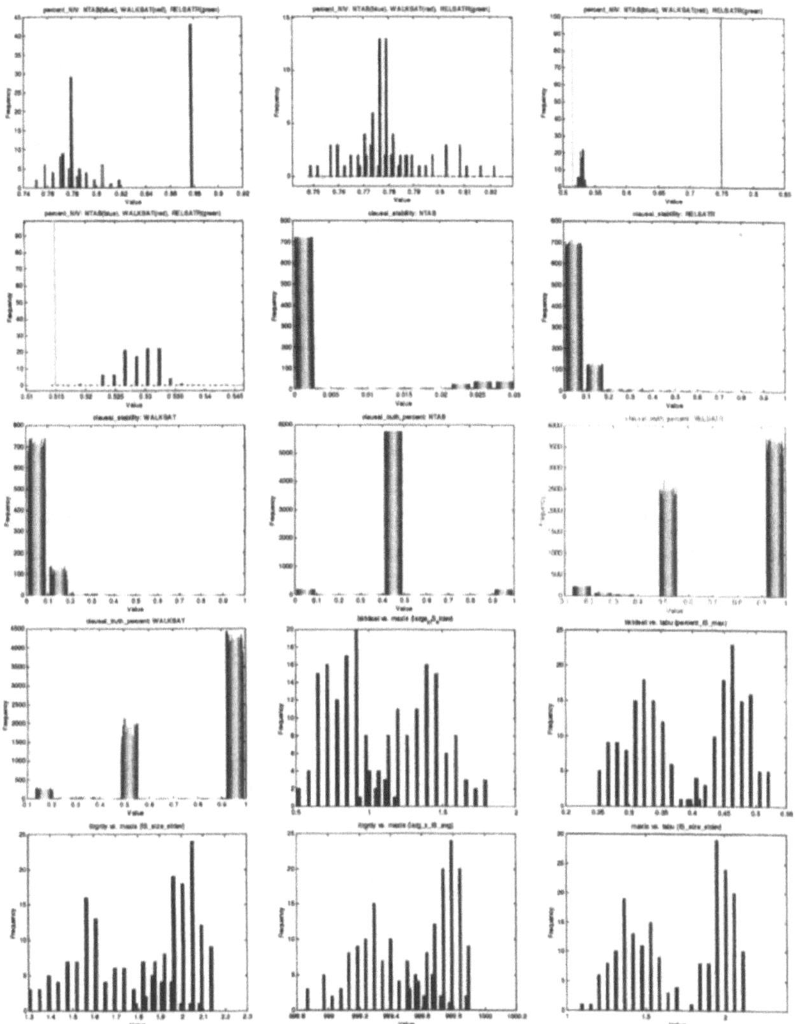

Figure 5.6. Each subfigure represents the following comparison (from upper left to bottom right): (1,3) π_1 and NTAB, Rel_SAT, and WalkSAT and (2,4) then zoomed version of the same property with only Rel_SAT, and WalkSAT, (5,6,7) π_2 for NTAB, Rel_SAT, and WalkSAT, and (8,9,10) π_3 for NTAB, Rel_SAT, and WalkSAT respectively. The last five subfigures depict the histograms of property value distribution for the following pairs of algorithms and properties: (11) DSATUR with backtracking vs. maxis and π_3, (12) DSATUR with backtracking vs. tabu search and π_7, (13,14) iterative greedy vs. maxis and π_1 and π_4, and (15) maxis vs. tabu and π_1.

backtracking vs. tabu search and π_7, iterative greedy vs. maxis and π_1 and π_4, and maxis vs. tabu and π_1 respectively.

Each of the diagrams can be used to associate a particular solution with one of the two algorithms A_1 and A_2 with 1% accuracy (100 instances attempted for statistics collection). For a given property value $\omega_i^{A_j} = x, j = 1, 2$ (x-axis), a test instance can be associated to algorithm A_1 with likelihood equal to the ratio of the pdf values (y-axis) $z(A_1, A_2)$. For the complete set of instances and algorithms that we have explored, as it can be observed from the diagrams, on the average, we have succeeded to associate 99% of solution instances with their corresponding algorithms with probability greater than 0.95. In one half of the cases, we have achieved association likelihood better than $1 - 10^{-6}$.

The forensic analysis techniques, that we have developed for solutions to SAT instances, have been tested using a real-life (circuit testing) and an abstract benchmark set of instances adopted from [Kam93, Tsu93]. Parts of the collected statistics are presented in the first ten subfigures in Figure 5.6. The subfigures represent the following comparisons: π_1 and NTAB, Rel_SAT, and WalkSAT and then zoomed version of the same property with only Rel_SAT, and WalkSAT (for two different sets of instances - total: first four subfigures), π_2 for NTAB, Rel_SAT, and WalkSAT, and π_3 for NTAB, Rel_SAT, and WalkSAT respectively.

The diagrams clearly indicate that solutions provided by NTAB can be easily distinguished from solutions provided by the other two algorithms using any of the three properties. However, solutions provided by Rel_SAT, and WalkSAT appear to be similar in structure (which is expected because they both use GSAT as the heuristic guidance for their prepositional search). We have succeeded to differentiate their solutions on per instance basis. For example, in the second subfigure it can be noticed that solutions provided by Rel_SAT have much wider range for π_1 and therefore, according to the second subfigure, approximately 50% of its solutions can be easily distinguished from WalkSAT's solutions with high probability. Significantly better results were obtained using another set of structurally different instances where among 100 solution instances no overlap in the value of property π_1 was detected for Rel_SAT, and WalkSAT.

GC Solvers	bktdsat	maxis	tabu	itrgrdy
bkdsat	998	2	0	0
maxis	3	993	0	4
tabu	1	0	995	4
itrgrdy	1	2	0	997

Table 5.4. Experimental Results: Graph Coloring. A thousand test cases were used. Statistics for each solver were established. The thousand instances were then classified using these statistics.

SAT Solvers	WalkSAT	RelSATR	NTAB
WalkSAT	992	5	3
RelSATR	6	990	4
NTAB	0	2	998

Table 5.5. Experimental Results: Boolean Satisfiability A thousand test cases were used. A thousand test cases were used. Statistics for each solver were established. The thousand instances were then classified using these statistics.

Using statistical methods, we obtained Table 5.4 and 5.5. A thousand test cases were classified using the statistical data. The rows of the tables represent the solver in which the thousand test cases originated from. The columns represent the classification of the solution using the statistical methods. In all cases more than 99% of the solutions were classified according to their original solvers with probability higher than 0.95. The Graph Coloring algorithms differ in many of the features, which resulted in very little overlap in the statistics. In the case of Boolean Satisfiability, both WalkSAT and Rel_SAT_rand are based on the GSAT algorithm which accounts for the slightly higher numbers when classifying between the two algorithms.

4. Public Detectable Watermarking Techniques

4.1 Introduction

Clearly the success of digital signatures relies on the detectability and traceability of the copyright marks. However, the general copy detection process is equivalent to the problems of pattern matching or subgraph isomorphism which are well-known NP-hard[61, 75, 88].

The pattern matching based technique chooses signatures selectively and develop fast comparison schemes to detect such signatures. To enable such fast comparison algorithms, one has to first identify a common structural representation of IPs and what constitutes an element of the IP structure; secondly, one has to determine a means of calculating locally context dependent signatures for such elements. Although this approach is generic, it is not always easy to find such common structure and design fast and accurate pattern matching algorithms.

The forensic engineering technique identifies solutions generated by strategically different algorithms. Each algorithm needs to be run on a large number of instances to collect statistical data, then these algorithms are clustered based on the obtained solutions' properties. To detect which algorithm is applied to obtain a given solution, we simply check its properties based on which the algorithm clustering has been performed.

Charbon and Torunoglu[31] discuss copy detection under a design environment that involves IPs from multiply sources that requires IP providers to register their IPs in a trusted agent. Their approach consists of two phases. First, a compact signature is generated from every IP block independently and made public. Then the IP integration process is performed in a way such that one can extract the signatures from the final design. However, every IP provider must do phase one and deposit its signature into a "bank" maintained by a third party. And again, matching algorithms need to be developed to detect signatures from a circuit.

In the broader area of information hiding, most of the reported literatures on detection focus on how to extract and recover the embedded-data with the secret key from the stego-data[124]. There are only two existing approaches to make watermark publicly detectable. One is based on the so-called public-key watermarking, the other relies on zero-knowledge protocols.

Hartung and Girod [72] present an extension of spread-spectrum watermarking that enables public decoding and verification of the watermark. However, an attacker can also discover the original watermark from the author's public key and remove it easily. Although the author can still retrieve the private part of the watermark by his private key, the property of public authentication is no longer there. To make it even worse, the attacker may further embed his own public watermark and claim his authorship.

Craver [44] uses zero-knowledge protocols to make the watermarks public enough to be detected yet private enough not to be removable. In such schemes, interaction between the detector and the author is required and the detector will challenge the author similar problems, possibly many times, to establish a proof of the authorship. More discussion on proving authorship of digital content can be found in [1] where a general model for proof of ownership is proposed.

Efficient detection technique is an essential piece of the protection mechanism and is as important as watermarking techniques. Compared to watermarking and fingerprinting, we see the research on copy detection lack both in breadth and in depth. Due to the hardness of the detection problem in general, most of the existing watermarking and fingerprinting literature focus on how to make the marks more secure and leave copy detection as an open challenge problem[75, 88]. The trade-off is that, in most cases, the more secure watermarks or fingerprints are, the more difficult to detect them, even for the authors.

The lack of detection mechanisms may cause problems for both IP providers and buyers who obtain IPs from other brokers and distributors. On one hand, if IP providers cannot detect their digital signatures, such marks become useless and IP's copyright is lost. On the other hand, dishonest parties may illegally sell the reproduced IPs to innocent buyers at a much lower price, knowing that the end users are unable to tell the real source of the IP. Things become even worse in the latter scenario, since IP buyers usually do not possess the knowledge that IP providers have for copy detection or the required expertise for forensic engineering.

In the remaining part of this section on copy detection, we describe an enhanced constraint-based watermarking technique that enables the embedded copyright mark to be easily and publicly detectable yet robust against forgery. Let us first see the following motivational example.

A Motivational Example

Kirovski et al.[88] propose a watermarking technique to hide signatures during logic synthesis where the marks are in the form of a set of primary outputs which are not necessary to be primary in the original design. The selection of such output nodes corresponds uniquely to the designer's or tool developer's signature. Constraints are introduced to enforce the selected gates to be visible in the final technology mapping solution. Suppose there are 100000 gates in a design out of which 10000 nodes are visible (LUT or cell outputs), and 1000 visible nodes are selected based on designer's secret key and the encryption scheme being used. The authors argue that the possibility that others accidentally obtain exactly the same solution is 10^{-1000}. The strength of this watermark relies on the uniqueness of these 1000 nodes. Designer's secret key is necessary for watermark detection.

Now we illustrate how public detectability can be achieved with the same example using the same watermarking method.

1 select 160 nodes, $G_1, G_2, \cdots, G_{160}$, and make them public (the selection of such gates will be discussed later);

2 hash a 4-letter design company symbol (32 bits in ASCII) by one-way hash function such as MD5;

3 append the 128-bit hash result to the 32-bit company symbol to make a 160-bit string: $m_1 m_2 \cdots m_{160}$;

4 for each nodes G_i, make it visible if $m_i = 1$ and invisible otherwise. Suppose that half of them (80 nodes) are made visible.

5 select 920 more nodes other than the 160 public nodes based on designer's secret key. Enforce these nodes to be visible to embed private watermark.

Overall, the new scheme chooses 1000 (920+80) non-primary nodes to be visible, therefore it can achieve the same level of protection as the previous one. (In fact, the new watermark is stronger since 80 more nodes are forced to be invisible.) Moreover, the public part of such watermark can be easily revealed without designer's secret key to establish a proof of authorship. More specifically,

1 check the visibility of gates $G_1, G_2, \cdots, G_{160}$ in that order and construct a 160-bit string $m_1 m_2 \cdots m_{160}$ by setting $m_i = 1$ if G_i is visible and letting $m_i = 0$ otherwise.

2 pick the first 32 bits $m_1 m_2 \cdots m_{32}$, which is the hidden plain text message in ASCII;

3 hash the selected 32-bit and compare the 128-bit hash result with $m_{33} m_{34} \cdots m_{160}$. If they match, the authorship is established. A mismatch indicates a sign of piracy and further careful moves should be considered (e.g. checking the visibility of the 920 nodes that carry the private watermark).

Benefits from the New Approach

The proposed public-private watermarking technique provides a practical and effective solution to the copy detection problem. The core concept is to divide the watermark into two parts: the public part which is visible to the public, and the private part which is only visible for authorized people. Both public watermark and private watermark are in the form of addition design constraints. Their difference is that public watermark is embedded in designated locations with known method to guarantee public detectability, while private part is embedded in a secret way as in the traditional constraint-based watermark. We use cryptographic techniques for data integrity to deter any attempt of removing or modifying the public watermark. The separation of public watermark and private watermark provides the following advantages:

- It facilitates easy public copy detection. A relatively convincing authorship can be verified by end users without forensic experts, which in great extend deters illegal redistribution.

- A 100% credibility is achievable (from the public part only) if this method is adopted by all IP providers. Otherwise, they can further select private watermark to obtain the desired level of credibility.

- Public watermark is hard to forge because it is generated by data integrity technique and embedded in the design process of VLSI IPs.

- There is little extra performance overhead, over traditional watermarking methods, to gain easy and public detectability.

- The new technique is compatible with all existing watermarking/fingerprinting methods.

4.2 Public-Private Watermarking Technique

Watermarking and fingerprinting are indirect protection schemes in that they provide a deterrent to infringers by providing the ability to demonstrate ownership of an IP to its originator[105]. The most popular watermarking and fingerprinting techniques are based on the addition of a pseudo-random bitstream as design constraints[27, 75, 80, 88, 120]. Such watermarks and fingerprints are invisible and in general robust because these additional constraints are integrated with the original ones during the design and implementation[80, 97, 132]. However, to detect these marks, either complete knowledge of the IP[31, 82, 98] or expertise on forensic engineering[90] are required as we have reviewed in the introduction section. The proposed public-private watermarking technique is a direct extension of the above idea based on constraint manipulation. We add a public portion of the watermark to simplify the detection process. We explain the global approach of the public-private watermarking technique in this section and leave the details of public watermarking to next section.

4.2.1 Watermark Selection and Embedding

The proposed public-private watermark consists of two parts: public and private, which are selected separately. We inherit the private part of the watermark from the traditional digital watermark discussed in early works[27, 75, 80, 88, 120]. A typical watermark is a cryptographically strong pseudo-random bit stream created by crypto systems using designer's digital signature as the secret key. Figure 5.7(a) shows how to create such bit streams. We hash the plain text message and get a 128-bit or higher hash result, which is used next as the key for a stream cipher to make the plain text message pseudo random.

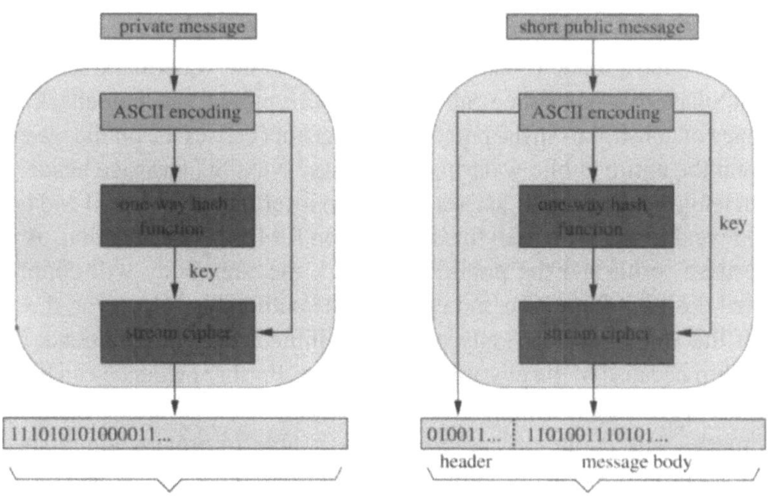

(a) private watermark message (b) public watermark message

Figure 5.7. Constructing public-private watermark messages.

The public watermark has a header and a body as shown in the bottom of Figure 5.7(b). To create such public watermark, we start with a short plain text message containing design information such as ownership, project title, or starting date. A good example may be the 4- or 3-letter symbol for the design company. The ASCII code of this short text is used as the public watermark header. We then use a one-way hash function to hash this ASCII code. The hash result is put into a stream cipher using the plain text message as the key. The output from the stream cipher is a pseudo-random bitstream and is appended to the watermark header as the body of the public watermark message.

Watermark embedding is the process of translating the binary watermark messages into design constraints. The public and private watermark can be embedded using either the same encoding scheme or different ones. The development of such schemes requires to explore the characteristics of the given problem and we will discuss this later (in the section of "Validation and Experimental Results") on specific VLSI CAD problems. However, to ensure that the public watermark be publicly detectable, we must make the followings public: (i) the hash function being used in the construction of public watermark, (ii) the (public) watermark encoding scheme being used to create the constraints, and (iii) the place where we embed these (public watermark related) constraints. We keep the secret key out of the reach of public to make the private watermark secure[3].

4.2.2 Watermark Detection and Security

We limit our discussion to the detection of public watermark, the private part can be detected by the existing copy detection techniques with the secret key [31, 82, 90].

Since we have made (i) the hash function, (ii) the watermark scheme, and (iii) the place that hosts the (public) watermark public, one can check for the existence of constraints in the part of the design that carries the public watermark to reveal the entire public watermark message. Next the message header (with known length) is extracted. This is the design information in ASCII and one can easily reveal it. One can hash this information for further verification. A match of the hash result and the public watermark message body will confirm the detected design information and establish the authorship. However, the public part of the watermark can only provide a limited level of confidence on the authorship because of the possibility of forgery. Further evidence can be shown when the secret key is available or forensic tools are used to detect the private watermark. The public-private watermark is able to provide simultaneously credibility as high as any traditional constraint-based watermark and the public detectability that no other watermarks can.

The private watermark is as secure as before, however the public part is visible to everyone and may be vulnerable against attacks. In most known constraint-based watermarking techniques, attackers will have a great amount of advantage if they can detect the watermark. This is not the case in our scheme because the message body in the public watermark is the hash of the message header. A perfect forgery will be the one that replaces the original public watermark by the adversary's public information. The adversary can follow the same steps to form his own public watermark using ASCII and the published hash function. It is trivial to identify the difference between this faked public watermark and the original one obtained by the above detection method. He then can alter the constraints based on the public-known watermark encoding scheme and embed them in the specific places. Finally, he must modify the known solution to satisfy these constraints. Such attack is possible, but it is hard and unrealistic[4] due to the following two facts:

- The faked hash will be different from the original in half of the bits statistically even if the message header is changed only by one bit. Therefore, we can make message body long such that this change will be significant.

- Design is an integrated process, it is unlikely one can make one change without altering the behavior of the design. At least some level of local modification is expected.

Another concern is how the public detectability of the public part affects the security of the private watermark. Since the selection and embedding of private watermark can be almost independent of the public watermark[5]. What the attacker gains from the public watermark are: the locations that we embed the public watermark which is negligibly small comparing to the entire design place, and the public watermarking scheme which may not be the same as the private watermarking scheme. These give attackers little help to break the private watermark.

4.2.3 Example: Graph Partitioning

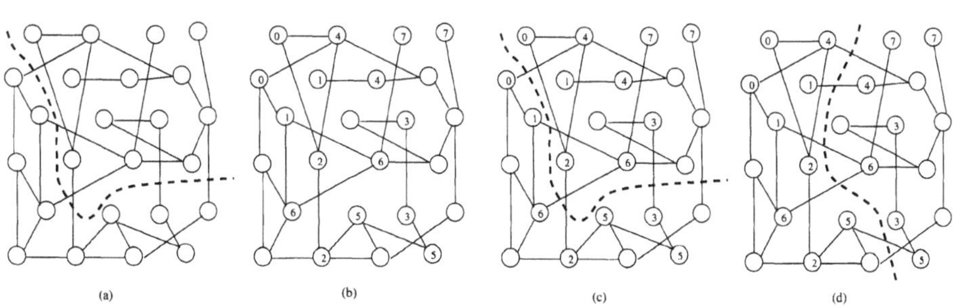

Figure 5.8. Public watermark on graph partitioning problem. (a) The original graph partitioning instance; (b) the same graph with 8 marked pairs that enables an 8-bit keyless public watermark; (c) A solution with public information "01001111"; and (d) A solution with public information "01110000".

Partitioning, which enables the powerful divide and conquer approach, plays a key role in VLSI design. We use this problem as an example to explain the basic concepts of the public watermarking. Given a hypergraph $G = (V, E)$ on a set of vertices V and a set of hyperedges E, the partitioning problem is to partition V into disjoint nonempty subsets. The constraints and objective functions for the partitioning vary with the level at which partitioning is performed and different design styles being used. Typical objective functions include minimizing interconnections and delay under constraints such as number of nodes in each partition (balance constraint), area of each partition and number of partitions.

Here we assume that we want to partition the graph in two subsets such that the number of edges being cut is minimized and the difference between the number of the two subsets is within two. For example, the dashed line in Figure 5.8(a) partitions the 24-vertex graph into two subsets, the one on its left contains 11 vertices and the other subset has 13. It cuts through 6 edges.

A k-bit public watermark is hidden in a graph partitioning solution as follows: we select k pairs of vertices and order them randomly; for each pair, we enforce them to be in different subsets to embed a bit 1 and enforce them to be in the same partition to embed a bit 0 by adding proper constraints. This is the so-called encoding scheme and the type of constraints depends on the objective function of the partition. For example, if we want to minimize the interconnection cost in a weighted graph, two vertices will go to different partitions if we change the weight of the edge between them to $-\infty$ (when they are connected) or add an extra edge of weight $-\infty$; similarly, they will stay together if the edge between them has a ∞ weight.

Figure 5.8(b) shows the 8 pairs of nodes that are picked to hold a 8-bit public watermark[6]. Figures 5.8(c) and (d) give two public watermarked solutions. The detection of these watermark is trivial. For example, the two vertices of pairs 0,1,2,3, and 6 in Figure 5.8(c) are separated, which implies 1's at the corresponding bit positions. The message has bit 0 at the other positions and we obtain the 8-bit message “01001111",which is ‘O’ in ASCII code. One can easily verify that the solution in Figure 5.8(d) hides the bit stream “0111000", i.e. letter ‘p’.

4.3 Theory of Public Watermarking

We elaborate in this section how to locate the positions for public watermark, how to create, embed, and detect such watermark.

4.3.1 General Approach

Figure 5.9 illustrates the generic public watermarking technique. We start with finding places in the original problem, which we call *public watermark holder*, to accommodate the public watermark. We then make the original prob-

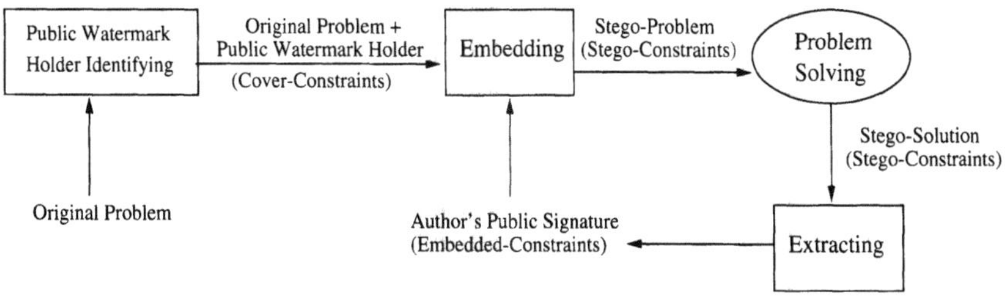

Figure 5.9. General approach of the public watermarking technique.

lem public with the identified public watermark holder as cover-constraints. The embedded-constraints corresponding with author's public signature will be added into the original problem in the public watermark holder. Solving the resulting stego-problem gives us a stego-solution that satisfies both the cover-constraints and the stego-constraints. The public watermark authentication is done in the extracting box, where one checks the satisfiability of the cover-constraints in the public watermark holder. Based on which constraint is satisfied, author's public signature can be retrieved from the known public watermark embedding scheme.

Embedding the author's public signature into public watermark holder with a known encoding watermarking scheme is unique for our approach. Although this is against the basic assumption that watermark should be "invisible" in the conventional constraint-based watermarking method[30, 75, 80, 133], it is the crucial step that enables public detectability. First, by hiding the public signature only at the known public watermark holder, instead of spread out all over the original problem, it becomes possible and inexpensive for everyone to know **where** to check for the public watermark. Secondly, unlike most information hiding techniques[124] and the earlier constraint-based watermarking method [80, 98, 164], we do not use any secret (stego-)key in the watermark embedding process. Furthermore, we make the encoding scheme public. Therefore, everyone, including IP buyers, will know **how** to extract the author's public signature.

4.3.2 Public Watermark Holder

We embed the public watermark by adding a special type of constraints: mutual exclusive constraints. We introduce the necessary definitions and explain them by the example of graph partitioning problem we discussed in the previous section Suppose that we want to partition a graph with k vertices, $\{v_1, v_2, \cdots, v_k\}$ into four subsets: $\mathcal{A}, \mathcal{B}, \mathcal{C}$ and $\mathcal{D}$.

Definition 1 (mutual exclusive):

Given a problem $\mathcal{P}$, a set of $n \geq 2$ constraints $\{C_1, C_2, \cdots, C_n\}$ are *mutual exclusive* if any solution $\mathcal{S}$ satisfies at most one constraint C_i, $(1 \leq i \leq n)$.

For example, the following four constraints are mutual exclusive for vertex v_1:

$C_1 : v_1$ must be in partition $\mathcal{A}$;
$C_2 : v_1$ must be in partition $\mathcal{B}$;
$C_3 : v_1$ must be in partition $\mathcal{C}$;
$C_4 : v_1$ must be in partition $\mathcal{D}$.

However, adding another constraint $\{C_1' : v_2$ must be in partition $\mathcal{A}\}$ makes the set of constraints $\{C_1, C_2, C_3, C_4, C_1'\}$ not mutual exclusive because a solution which places both v_1 and v_2 in subset $\mathcal{A}$ will satisfy both C_1 and C_1'.

Definition 2 (complete mutual exclusive set):

A mutual exclusive set of constraint $\{C_1, C_2, \cdots, C_n\}$ is *complete* if any solution $\mathcal{S}$ satisfies exactly one constraint.

The set $\{C_1, C_2, C_3\}$ is mutual exclusive, but not complete because any solution that has vertex v_1 in partition $\mathcal{D}$ will not satisfy any of these three constraints. Adding constraint C_4 makes it complete.

Definition 3 (strongly mutual exclusive set):

A mutual exclusive set is *strongly mutual exclusive* if for any constraint C_i, there exists a solution $\mathcal{S}$ that satisfies C_i and violates $C_j (j \neq i)$.

For any constraint C_i in the mutual exclusive set $\{C_1, C_2, C_3, C_4\}$, we can first fix v_1 in the subset that corresponds to C_i, then partition the rest vertices. This gives us a solution that satisfies only C_i and violates all the other constraints. Therefore, it is a strongly mutual exclusive set.

Theorem 1 (Existence Theorem):

Complete strongly mutual exclusive set exists for all problems with more than two different solutions.

[Proof:]

Suppose S and S' be two different solutions to a given problem, then we can always find one property that S satisfies but S' does not. This is because of the distinctness of S and S'. Denote this property by $\mathcal{P}$ and define the following two constraints:

$C_1 :$ a solution must have property $\mathcal{P}$
$C_2 :$ a solution should not possess property $\mathcal{P}$

It is easy to see that any solution satisfies exactly one of these two constraints, moreover, solution S meets constraint C_1 and S' meets C_2. Therefore the set $\{C_1, C_2\}$ is a complete strongly mutual exclusive set for the given problem. □

Theorem 2 (Cutting Space Theorem):

A set of complete strongly mutual exclusive constraints partitions the solution space as the union of nonempty disjoint subsets.

[Proof:]

Let $\{C_1, C_2, \cdots, \}$ be the set of complete strongly mutual exclusive constraints and $\mathcal{S}$ be the set of all solutions to a given problem. Define $\mathcal{S}_i\mathcal{S}$ be the set of all solutions that satisfy constraint C_i. $\mathcal{S}_i \neq \phi$ because the set of constraints is strong. The mutual exclusiveness of the constraints implies that $S_i S_j = \phi$ for all $i \neq j$. Finally, $S_i = S$ since the set is complete, i.e. any solution must satisfy one of the constraints. □

Now suppose that we have a set of complete strongly mutual exclusive constraints $\{C_1, C_2, \cdots, C_n\}$[7], if one chooses constraint C_i as his public watermark holder, then he can only find solutions from the subset of solutions satisfying C_i. From the cutting space theorem, we conclude that there will not be any collision[8] between any two with different watermark holders. This essentially provides the ultimate 100% proof of the authorship. Furthermore, this is independent of the length of the public watermark. In sum, we have:

Theorem 3 (Data Hiding Theorem):

n different pieces of information (of any length) can be hidden with a (complete) strongly mutual exclusive set of n constraints.

Therefore, it is of our interest to find large complete strongly mutual exclusive set to accommodate the possible large number of public watermarks. We now introduce the concept of *join* and explain a systematic method to construct complete mutual exclusive sets.

Definition 4 (join):

The *join* of two sets of constraints, $\{C_1, \cdots, C_n\}$ and $\{C'_1, \cdots, C'_m\}$, is defined as the set $\{C_1 \wedge C'_1, \cdots, C_1 \wedge C'_m, C_2 \wedge C'_1, \cdots, C_n \wedge C'_m\}$, where constraint $C_i \wedge C'_j$ is satisfied if and only if both constraints C_i and C'_j are satisfied.

The number of constraints grows rapidly with the join operation. Moreover, join preserves the mutual exclusiveness and completeness. (I.e., the join of two complete mutual exclusive sets is also a complete mutual exclusive set.). However, it does not guarantees the strongly mutual exclusiveness required by the data hiding theorem. For example, the set

$C_1 : v_1$ and v_2 in partition $\mathcal{A}$;
$C_2 : v_1$ in partition $\mathcal{A}, v_2$ not in partition $\mathcal{A}$;
$C_3 : v_1$ not in partition $\mathcal{A}, v_2$ in partition $\mathcal{A}$;
$C_4 : v_1$ and v_2 not in partition $\mathcal{A}$;

and the set

$C'_1 : v_1$ and v_2 in the same partition;
$C'_2 : v_1$ and v_2 in different partitions;

are both complete strongly mutual exclusive sets. But there does not exist any solution satisfying the constraint $\{C_{12} : v_1$ and v_2 in partition $\mathcal{A}$ and v_1 and v_2 in different partitions.$\}$.

Failing to preserve the strongly mutual exclusiveness makes join improper for creating large strongly mutual exclusive set of constraints. Because join may introduce constraints that no solution can satisfy. Therefore if one's signature is mapped to such constraint, he will not be able to solve the problem. We observe that this is caused by the dependency of constraints in different sets.

For example, the natural conflict between C_1 and C_2' implies that their join C_{12} cannot be satisfied by any solution. We introduce the concept of independent constraints to solve the problem.

Definition 5 (independent constraints):

Two constraints are *independent* if any solution's satisfiability to one constraint has no impact on its satisfiability to the other one. Two sets of constraints, $\{C_1, C_2, \cdots, C_n\}$ and $\{C_1', C_2', \cdots, C_m'\}$, are *independent* if C_i and C_j' are independent for any $1 \leq i \leq n$, and $1 \leq j \leq m$.

Theorem 4 (Join Theorem):

If two complete strong mutual exclusive sets are independent and have n and m constraints respectively then their join is a complete strong mutual exclusive set with $n \cdot m$ constraints.

[Proof:]

Let $C = \{C_1 \wedge C_1', \cdots, C_1 \wedge C_m', C_2 \wedge C_1', \cdots, C_n \wedge C_m'\}$ be the join of two complete strong mutual exclusive sets $\{C_1, \cdots, C_n\}$ and $\{C_1', \cdots, C_m'\}$. We first show that C is a complete strong mutual exclusive set.

If one solution satisfies two constraints $C_i \wedge C_j'$ and $C_{i'} \wedge C_{j'}'$, then it satisfies all the four constraints $C_i, C_{i'}, C_j'$, and $C_{j'}'$ from the definition of join. Since the sets $\{C_1, \cdots, C_n\}$ and $\{C_1', \cdots, C_m'\}$ are mutual exclusive, we have $i = i'$, and $j = j'$. Therefore, $C_i \wedge C_j' = C_{i'} \wedge C_{j'}'$ and C is mutual exclusive.

For a given solution, let C_i and C_j' be the constraints it satisfies. The completeness of sets $\{C_1, \cdots, C_n\}$ and $\{C_1', \cdots, C_m'\}$ guarantees the existence of such C_i and C_j'. From the definition of join, we know this given solution meet the constraint $C_i \wedge C_j'$ of set C. So C is complete.

For any constraint $C_i \wedge C_j'$, there exist solutions that satisfy either C_i or C_j' because the original constraint sets are strong. If there is no solution that meets both C_i and C_j', then all the solutions that satisfy C_i will not meet C_j', and vice versa. Or in another word, satisfying one constraints prevents the satisfaction to another. The contradiction to the independence of C_i and C_j' implies that the mutual exclusive set C is strong.

Clearly C consists of $n \cdot m$ constraints from the way it is constructed, we only need to show that these constraints are all distinct. Suppose that $C_i \wedge C_j' = C_{i'} \wedge C_{j'}'$ but $i \neq i'$ or $j \neq j'$. Denote $P = C_i$
$C_{i'}$ be the property (constraint) required by C_i, but not by $C_{i'}$. Similarly let $Q = C_{i'}\ C_i$. Now we consider the constraint $P \wedge Q$. Apparently $P \wedge Q$ is weaker[9] than $C_i \wedge C_j' = C_{i'} \wedge C_{j'}'$, thus any solution that satisfies $C_i \wedge C_j'$ will satisfy $P \wedge Q$ as well. The set $\{C_1, \cdots, C_n\}$ is complete mutual exclusive, so there exist one and only one k such that C_k is stronger than $P \wedge Q$. For the same reason, we can find a unique constraint $C_{k'}'$ from $\{C_1', \cdots, C_m'\}$ that is stronger than $P \wedge Q$. For any solution that satisfies C_k, $P \wedge Q$ is also satisfied. Since $C_{k'}'$ is the only constraint from $\{C_1', \cdots, C_m'\}$ that is stronger

than $P \wedge Q$, this solution cannot satisfy any other constraints. Furthermore, $C'_{k'}$ is actually satisfied because of the completeness of the set. In sum, any solution that satisfies C_k satisfies $C'_{k'}$ and vice versa. That is $C_k = C'_{k'}$, which is a contradiction to the independence between the two sets. □

4.3.3 Public Watermark Embedding

We now explain the embedding box, the second step in public watermark embedding, in Figure 5.9. Its function is to create the stego-problem that corresponds to the author's public signature. It consists of three phases: constructing mutual exclusive set of constraints, creating public watermarks, and defining public watermark embedding schemes. We have built the theory on how to define public watermark holders (the mutual exclusive set of constraints) and discussed how to create and embed public watermark. We now complete the graph partitioning example to elaborate the entire process.

Construct Mutual Exclusive Set of Constraints

Suppose we want to partition a graph with n vertices, $\{v_1, \cdots, v_n\}$, into two partitions. We select $2k$ distinct vertices (for example randomly): $v_{i_1}, v_{i'_1}; v_{i_2}, v_{i'_2}; \cdots; v_{i_k}, v_{i'_k}$. Define k sets of constraints: $\{C_{i_1}, C'_{i_1}\}, \cdots, \{C_{i_k}, C'_{i_k}\}$ as

C_{i_j} : vertices v_{i_j} and $v_{i'_j}$ are in the same partition.

C'_{i_j} : vertices v_{i_j} and $v_{i'_j}$ are in different partitions.

It is easy to verify that every set is complete strongly mutual exclusive and these k disjoint sets are independent. The join of these k sets gives us a complete strongly mutual exclusive set with 2^k constraints:

$$\{\bar{C}_{i_1} \wedge \bar{C}_{i_2} \wedge \cdots \wedge \bar{C}_{i_k} : \bar{C}_{i_j} = C_{i_j} \text{ or } C'_{i_j}\} \qquad (*)$$

where the join constraint $\bar{C}_{i_1} \wedge \cdots \wedge \bar{C}_{i_k}$ is satisfied if and only if all $\bar{C}_{i_j}$'s are satisfied. For example, $C_{i_1} \wedge C_{i_2} \wedge \cdots \wedge C_{i_k}$ is the constraint that requires vertices v_{i_j} and $v_{i'_j}$ be in the same partition for all $1 \le j \le k$.

Create Public Watermarks

Figure 5.10 shows step-by-step how to create the keyless public watermark from author's public signature. The public watermark is a bitstream with a header and a body. The header is just the author's plain text public signature (with a fixed length) in ASCII code. This ASCII code is hashed by a one-way hash function (e.g. MD5); the hash is put into a stream cipher (e.g. RC4) with the ASCII code as key and the produced pseudo random bitstream makes the body of the public watermark. The simplicity of watermark header facilitates public authentication and the pseudo-random watermark body provides robustness again attacks which we will discuss in the section of authentication.

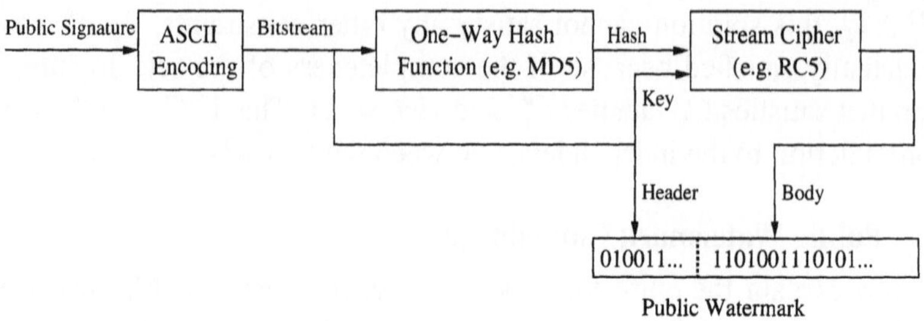

Figure 5.10. Creating keyless public watermark from public signature.

Define Embedding Schemes

Now we have a set of mutual exclusive constraints and a set of public watermarks. A *watermark embedding scheme* is a one-to-one function from the set of watermarks to the set of constraints such that different public watermarks are mapped to different constraints. We intend to keep the embedding scheme as simple as possible for the purpose of public authentication.

As a continuation of the previous graph partitioning example, we can define the watermark embedding scheme as follows:

> *for public watermark* $p_1p_2\cdots p_k$, *we choose the constraint* $\bar{C}_{i_1}\wedge\cdots\wedge\bar{C}_{i_k}$ *from the complete strongly mutual exclusive set* $(*)$ *we constructed earlier, where* $\bar{C}_{i_j}=C_{i_j}$ *if* $p_j=0$ *and* $\bar{C}_j=C'_{i_j}$ *if* $p_j=1$.

The stego-problem is obtained by adding *THE* constraint that corresponds to the public watermark under the embedding scheme. Different watermarks are mapped to different constraints from a strongly mutual exclusive set. Therefore all stego-problems will be different and the property of mutual exclusiveness guarantees their solutions will be distinct. In sum, we have

Theorem 5 (Correctness of the Approach):

If the constraints are strongly mutual exclusive, there always exist (stego-) solutions for the stego-problem. Furthermore, different stego-problems will have different (stego-)solutions which are all solutions to the original problem.

4.3.4 Public Watermark Authentication

In this part, we explain the extracting box in Figure 5.9 whose function is to detect the public watermark from a given stego-solution and retrieve the author's public signature. The followings are available to the public: **(i)** the original problem; **(ii)** the set of mutual exclusive constraints which is the public watermark holder; **(iii)** the public watermark embedding scheme; **(iv)** the fixed length of all author's public signature; and **(v)** a stego-solution needs for authentication.

A detector checks which constraint from the set of mutual exclusive constraints **(ii)** does the given stego-solution **(v)** satisfy. Then he obtains the embedded public watermark from the known embedding scheme **(iii)**. He now takes the watermark header of fixed length **(iv)**, this gives the author's public watermark in ASCII format and suggests the possible author. The detector may further hash this watermark header and use the stream cipher to re-produce the watermark body. A strong proof to the authorship is established if the re-produce the watermark body coincides with the one extracted from the stego-solution.

4.3.5 Summary

We summarize this section with the following remarks on the public-private watermark:

Credibility: The public watermark gives a perfect proof to the authorship. The mutual exclusiveness guarantees different stego-solutions for distinct public signatures.

Public watermark header: This is the key that enables the watermark to be detected publicly. It is important to keep it in plain text for the authentication purpose.

Public watermark body: This is the part that secures the public watermark. For many problem, one may find a new solution based on the given solution by study the locality of the problem. With only a short header, the public watermark is vulnerable to forgery. The public watermark body provides the public watermark integrity and makes forgery hard (theoretically, even one bit change in watermark header results in half of the bits being flipped in the watermark body).

Join: The join operation provides an efficient way to produce large set of mutual exclusive constraints. It also enables a logarithmic time IP authentication instead of linear.

Impact to the quality of the solution: Similar to the conventional constraint-based watermarking techniques, adding extra constraints may introduce degradation of the solution's quality. One of the criteria for building mutual exclusive constraints is to keep this overhead at the minimum level.

Robustness: The stego-solution is obtained by solving the stego-problem which contains a unique public watermark. A successful forgery is a different solution obtained by modifying the given solution and has the attacker's public watermark embedded. A different solution may not be difficult to get. However, it is hard in general to hide another information unless the attacker is able to solve the problem by himself in which case he has little incentive for forgery.

Public-private watermarking technique: The public watermark technique is compatible with all the existing watermarking techniques. The proposed public-private watermarking approach allows authors to embed more information based on their secret keys after the public watermarks are enforced. It is used to enhance the watermark's credibility.

4.4 Validation and Experimental Results

We have explained how to create the public-private watermark which is a pseudo-random bit stream (except the header of public watermark). In this section, we conduct case studies on several well-known VLSI CAD problems to validate this approach. First, we give a tangible example of public watermark on the graph partitioning problem. Then we show to combine this with an existing FPGA watermarking technique to achieve the public detectability. We demonstrate the robustness of the public watermark via the Boolean satisfiability problem and finally discuss its impact to system's performance within the context of graph coloring.

4.4.1 FPGA Layout

Lach et al.[97] propose an FPGA fingerprinting technique that utilize the FPGA design flexibility to put a unique identification mark into the design for each customer. For example, the four tiles in Figure 5.11, each contains four configurable logic blocks, all implement the same Boolean function $Z = A + B + C \cdot D$. Moreover, they have the same interfaces and thus are interchangeable. The timing of the circuit may vary due to the changes in routing.

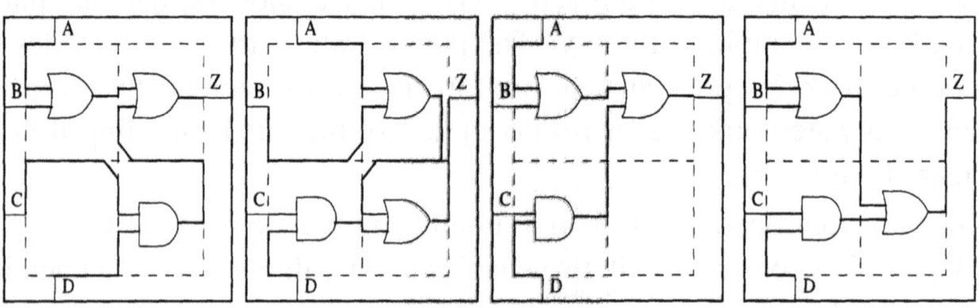

Figure 5.11. Four instances of the same function with fixed interfaces (redrawn from [97]).

This observation is used to create different design for different customer to trace the use of the design[97]. However, the same property can be used to embed public watermark. We first label the four CLBs as 00, 01, 10, 11 clockwise from the upper left to the lower left. To hide 2 bits from the public watermark message, one can choose one of these four implementations, with the unused CLB has the same label as the given 2 bits. For example, from left to right, the four design in Figure 5.11 have "11", "00", "10", and "01" as the embedded message respectively. With a few of such tiles, one can find sufficient space for public watermark messages.

Forgery is a problem for this approach. Given a FPGA layout with the public-private watermark embedded, an attacker can go to the tiles where public watermark is hidden and obtain the bit stream easily. Then he can change the message header at his wish, use one-way hash function and stream cipher on his new message header to forge a message. Next, he can do the necessary modifications in these tiles to replace the original public watermark by his faked message. This will be a successful attack unless private watermark is revealed. However, this is the same problem as what FPGA watermarking and fingerprinting techniques are facing. The solution lies on the difficulty of reverse engineering and the fact that most FPGA vendors will not reveal the specification of their configuration streams[80, 97].

4.4.2 Boolean Satisfiability

The Boolean satisfiability problem (SAT) seeks to decide, for a given formula, whether there is a truth assignment for its variables that makes the formula true. SAT appears in many contexts in the field of VLSI CAD, such as automatic pattern generation, logic verification, timing analysis, delay fault testing and channel routing. We necessarily assume that the SAT instance to be protected is satisfiable and that there is a large enough solution space to accommodate the watermark.

Given a formula $\mathcal{F}$ on a set of boolean variables V, the simplest watermarking technique for public detectability is to hide the public watermark behind a known subset of variables $\{v_1, v_2, \cdots, v_k\}$. Suppose the public watermark message is $m_k \cdots m_2 m_1$, we embed it by forcing $v_i = m_i$ in the solution. This can be done by adding to the formula $\mathcal{F}$ single-literal clause v_i (if $m_i = 1$) or v_i' (if $m_i = 0$)[10].

We pick four 4-letter messages A, B, C, and D. We use MD5[138] (source code available at *ftp://ftp.sunet.se/pub3/vendor/sco/skunkware/uw7/fileutil/md5/src*) as the one-way hash function to obtain four 128-bit messages H(A), H(B), H(C), and H(D). Next we use RC4 (source code available at *ftp://ftp.ox.ac.uk/pub/crypto/misc/rc4.tar.gz*) to encrypt these messages using their ASCII codes as the encryption keys. The resulting pseudo-random bit streams are appended to the ASCII codes of the corresponding plain text to form the four public watermark messages as illustrated in Figure 5.7(b).

Figure 5.12 shows pairwisely the Hamming distance among these four public watermark message. A and B, B and D are relatively close because each pair has one letter in common accidentally[11].

We now embed these public watermark messages to DIMACS SAT benchmarks, where the instances are generated from the problem of inferring the logic in an 8-input, 1-output "blackbox" (*http:// dimacs.rutgers.edu/*). We first select 32 variables for the message header, then choose 128 (or 64 for instances of small size, e.g., with less than 600 variables) more variables for the message body. We then assign values to these variables based on the public watermark and solve for the assignment of the rest variables to get the original solution.

With the given solution (and variables that carry the public watermark), an adversary retrieve the public message header, modify it and compute the new

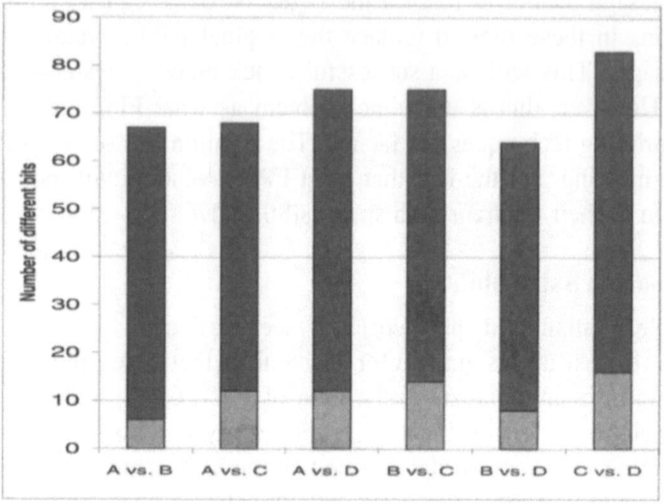

Figure 5.12. Hamming distance among the four public watermark messages. The bottom half comes from the message header(plain text part), and the top half comes from the message body(results of RC4).

		4 bits in header		8 bits in header		16 bits in header		24 bits in header	
$\mathcal{F}$	N	body	sol.	body	sol.	body	sol.	body	sol.
ii8b1	336	31.2	148	32.8	150	31.8	168	32.6	170
ii8b2	576	33.6	260	30.6	258	32.4	265	32.0	272
ii8b3	816	62.2	363	64.0	376	67.4	358	61.6	387
ii8b4	1068	65.8	489	66.2	472	63.4	492	62.6	513
Ave. Dist. (%)		40.2%	-	43.5%	-	50.5%	-	56.2%	-
Ave. Dist. (%)		-	44.9%	-	44.9%	-	46.5%	-	48.3%

Table 5.6. Average number of different bits in public message body ("body"), average distance (rounded to integer) from the original solution ("sol.") when 4-bit, 8-bit, 16-bit, and 32-bit forgery is conducted to the public message header on SAT benchmarks.

message body. He then embed this forged message and resolve the problem. Our goal is to show that there is little correlation between the original solution and adversary's new solution, i.e., attacker has little advantage from the original solution or it is equally difficult to obtain a solution.

Table 5.6 shows our experimental results, where messages A, B, C, D are embedded to the four SAT instances respectively. The second column gives the number of variables N in these instances. We consider the adversary changes randomly 4 bits, 8 bits, 16 bits, and 24 bits in the 32-bit message header. We repeat each trial 5 times, the columns labeled "body" show the average number of bits changed in the faked message body from the original. We solve each instance with this faked message (both header and body) embedded and calculate the Hamming distance between the new solution and the original solution. The average distances (rounded to the nearest integer) are reported in columns with label "sol.".

The last two rows report these average distances percentage-wise. The first is the distance in public domain, which is very close to 50% if we exclude the mandatory header part. It is independent of the number of bits being modified in the header and shows the robustness of our cryptographic tools in generating pseudo-random bit streams. The last row shows that the new solutions are not close to the original solution. (When we solve the original instances for multiple solution, their average distance is also about 45%.) Therefore, we can conclude that the new solutions are independent of the given solution, which means that once the public watermark has been modified, the adversary loses almost all the advantage from the given solution. This is further verified by the fact that the run time difference for resolving the problem and solving from scratch is so small (within 5%) that we consider they are the same.

4.4.3 Graph Coloring

The NP-hard graph vertex coloring optimization seeks to color a given graph with as few colors as possible, such that no two adjacent vertices receive the same color. We propose the following public-private watermarking technique for graph coloring problem and use it to demonstrate our approach's impact to the quality of the solution:

> For a given graph, we select pairs of vertices that are not connected directly by an edge. We hide one bit of information behind each pair as follows: adding one edge between the two vertices and thus making them to be colored by different colors to embed 1; collapsing this pair and thus forcing them to receive the same color to embed 0.

Consider Figure 5.13, two pairs of unconnected vertices, nodes 0 and 7, and nodes 1 and 8, are selected as shown in the dashed circles in 5.13(a). The rest of Figure 5.13 shows four different coloring schemes with a 2-bit public watermark message embedded. To detect such watermark, one can simply check the colors received by nodes 0, 1, 7, and 8. For example, in Figure 5.13(c), nodes 0 and 7 are colored by G(reen) and Y(ellow) respectively, which means the first bit (the most significant bit) is 0. Similarly, the observation that nodes 1 and 8 are both colored by R(ed) tells us the second bit of the message is 1. Therefore, we detect a public message "01".

To evaluate the trade-off between protection and solution degradation (in the case of graph coloring, the number of extra colors), we first color the original graph, then color the watermarked graph and comparing the average number of colors required. We consider two classes of real life graphs (the *fpsol2* and

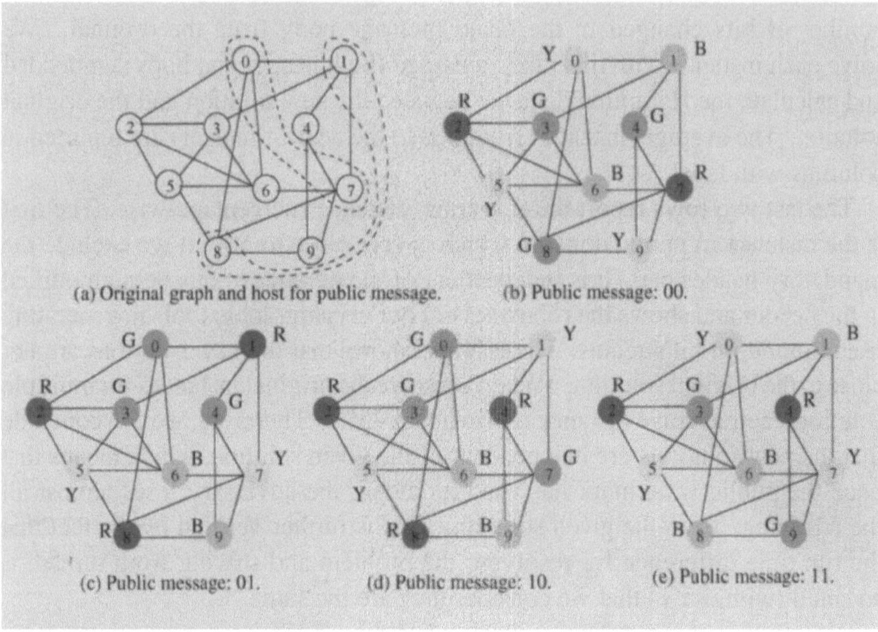

Figure 5.13. Four GC solutions with different public watermarks added to the same graph.

inithx instances from *http://mat.gsia.cmu.edu/COLOR/instances.html*) and the DIMACS on-line challenge graph (available at *http://dimacs.rutgers.edu/*).

original instance			32-bit message		64-bit message	
	vert.	opt.	overhead	best	overhead	best
fpsol2.i.1	496	65	0.2	65	0.7	65
fpsol2.i.2	451	30	0.1	30	0.5	30
fpsol2.i.3	425	30	0.1	30	0.5	30
inithx.i.1	864	54	0.0	54	0.2	54
inithx.i.2	645	31	0.9	31	1.8	32
inithx.i.3	621	31	1.1	31	1.9	32
DSJC1000	1000	85.8	0.5	86	2.0	87

Table 5.7. Embedding public watermark to real-life graphs and randomized graphs.

Table 5.7 shows the number of vertices in each graph, the optimal solutions (the DSJC1000 problem is still open. The number in the table is the average of 10 trials with 85-color solutions occur several times), and the overhead introduced by public watermark messages of various length. For each instance, we create ten 32-bit and ten 64-bit public watermark messages randomly. We add the message to the graph and color the modified graph. The average number of colors and the best solution we find are reported. One can easily see that the proposed approach causes little overhead for real life instances, but loses best solutions for the randomized DSJC1000 graph. The reason is that there exist localities in real life graph of which we can take advantage of. However, such localities do not exist or are very difficult to find in random graphs.

5. Summary

The goal of copy detection techniques is to discover the embedded copyright marks from the IP. Its importance to the constraint-based IP protection paradigm is needless to say, all watermarks and fingerprints are useless unless they can be accurately and effectively identified. However, the general copy detection problem is computational intractable. In this chapter, we discuss this challenging problem and propose three strategically different copy detection approaches.

The pattern matching based technique is most natural copy detection method, where the digitalized signature are carefully selected to facilitate fast comparison schemes (pattern matching algorithms) for detection. Its drawback is that, to enable the fast comparison algorithms, one has to first identify certain common structural representation of IPs and what constitutes an element of the IP structure. In addition, one has to determine how to calculate locally context dependent signatures.

The forensic engineering technique seeks to identify the source of a given piece of IP from a pool of IP sources. To detect which algorithm is applied to obtain a given solution, one simply needs to check a set of properties based on which the strategically different algorithms (IP sources) are clustered. The main difficulty for this approach is how to extract such properties and in general the detailed information of the algorithms is required.

The public detectable watermarking technique is an enhanced watermarking method that facilitates easy and public detection. This is achieved by allowing part of the watermark to be public. Cryptographic techniques, in particular techniques for data integrity, are used to protect the public watermark from forgery. Although this new approach is compatible with all the existing watermarking techniques and has the potential of solving eventually the IP protection problem, it needs help from industrial organizations to push for design standards.

Notes

1 The value of parameter N determines the sensitivity of the copy detection process. Larger values enable the algorithm to handle greater perturbations by instruction reordering, but increase runtime since more patterns are generated.

2 Even if some sequences are the same, this does not mean that the netlists are isomorphic. However, the procedure will leave only a few candidates for stolen IP fragments, and these can be checked in essentially linear time. Vertices can also be annotated with information (logic type, hierarchy level, etc.) to induce corresponding marked degree sequences, as discussed below – again, this is to produce a staged "filtering" before applying detailed isomorphism tests.

3 The security of the cryptographic function depends on the secret key, not on which hash function or stream cipher we use to encrypt the message. Also it is the digital signature, which is independent of the watermark encoding schemes, that carries the proof of authorship.

4 By *unrealistic*, we mean that the performance degradation of the modified IP is so large that one will not accept it and the design loses its value.

5 We say *almost independent* because the selection and embedding of private watermark are restricted by the existence of public watermark to certain extend. For example, the addition of private watermark should not change the public watermark.

6 Due to the small size of the example, we assume that we have only the public watermark message header here. The encrypted message body can be embedded and detected in the same way.

7 We can easily make a set of strongly mutual exclusive set $\{C_1, C_2, \cdots, C_{n-1}\}$ complete by adding the constraint C_n : "$C_1, C_2, \cdots, C_{n-1}$ all fail".

8 A collision occurs when one solution meets more than one public watermark. In such situation, one cannot identify the real author(s) and the watermark fails.

9 If all solutions that satisfy constraint C also satisfy constraint C', then we call C' is *weaker* than C and C is *stronger* than C'.

10 A single-literal clause imposes a very strong constraint to the formula. Statistically it will cut the entire solution space by one half. Therefore we may use a short public watermark message, in particular for instances with not so many variables. However, the credibility can always be enhanced by adding private watermark using other techniques, such as those proposed in [80].

11 The ASCII codes for messages A, B, C, and D are: "01010011 01000111 01001001 00100000", "01000011 01000100 01001110 00100000", "01010011 01001110 01010000 01010011", and "01001101 01000101 01001110 01010100".

Chapter 6

CONCLUSIONS

We have witnessed the thriving of embedded system in the past decade. The rapid development of silicon capacity, advances in fabrication technologies, and the emergence of the Internet and World Wide Web provide all the necessary condition for the network-centered embedded systems to explode. This imposes challenges in almost all areas of computer science and engineering: computer architecture, compilers, operating system, and so on. In particular, it has changed the system design philosophy. With the system-on-chip and new design objectives such as low cost, high performance, high portability, low power consumption, and short time-to-market, intellectual property (IP) reuse has emerged as a vital and growing business in semiconductor and system design industry. Traditional design methodology has stepped down to IP-based design.

Effective IP protection technique is one of the enabling technologies for industrial-strength IP-based synthesis. This book provides an overview of the security problems in modern VLSI design with a detailed treatment of our newly developed constraint-based protection paradigm that consists of watermarking, fingerprinting, and copy detection.

The goal of IP protection is to discourage or deter illegal IP copying and redistributing. If IP misappropriation occurs, the IP protection techniques should be able to help the IP provider in (i) proving the authorship and (ii) finding the dishonest IP user. Failure to either one of these will not give a true and complete protection for the IP.

The problem of VLSI design IP protection is much more challenging than the protection of artifact-type of IPs (text, image, audio, video, and multimedia contents) because design IPs are sensitive to errors and their correct functionality must be preserved. This difference prevents us from directly applying the state-of-the-art artifact protection techniques, most of which modify the contents to certain extend for protection, to design IP protection. Clearly the main difficulty is how to gain protection without rendering the IP useless.

The new constraint-based IP protection methods are based on the observation that (i) the design and implementation of IPs are similar to the process of problem solving, and (ii) there usually exists a large solution space. That is, IP development is a constraint satisfying problem where we optimize the design objectives subject to the design specifications (constraints). There are different solutions that meet all the constraints, corresponding to different design styles and implementation of the IP with the same functionality.

Our key idea is to superimpose additional constraints that correspond to an encrypted signature of the designer to design/software in such a way that quality of design is only nominally impacted, while strong proof of authorship is guaranteed. The addition of signature-related constraints restricts designers to a smaller subspace of the original solution space for IP implementation. Any reported solution from this smaller subspace will (i) be a valid solution to the original problem, and (ii) meet the additional design constraints that are not necessarily to be satisfied for a solution to the original problem. The first guarantees the value of the IP and the second provides evidence of designer's authorship.

The proposed constraint-based IP protection paradigm consists of three integrated parts: constraint-based watermarking, fingerprinting, and copy detection. Its correctness relies on the presence of all these components. In short, watermarking aims to embed signatures for the identification of the IP owner without altering the IP's functionality; fingerprinting seeks to provide effective ways to distinguish each individual IP users to protect legal customers; copy detection is the method to catch improper use of the IP and demonstrate IP's ownership.

Constraint-based watermarking technique encodes signature as additional constraints, adds them into the problem specification and solve this more constrained problem instead of the original problem. The authorship is proved statistically by showing the (usually extremely) small likelihood that a randomly found solution to the original problem satisfies all or most of the additional constraints. Besides keeping the IP's correct functionality, a good watermark should provide a high credibility, introduce low overhead, remain high resilient in the IP, be transparent to the IP design and implementation process, be perceptually invisible, and offer part protection.

The goal of fingerprinting is to protect innocent IP users whenever IP misuse or piracy occurs. It is clear that to enable this, assigning different users distinct copies of the IP (with their fingerprints embedded) becomes necessary. In addition to the above requirements for watermarks, we demand fingerprints (as the user's watermark) to have the following attributes: low runtime overhead, collusion-secure, high traceability, and preserving watermarks. Clearly the key challenge is how to produce efficient copies of fingerprinted IPs with little runtime overhead.

Copy detection is an important part of our constraint-based IP protection paradigm. It targets to find the IP provider's watermark and IP buyer's fingerprint in a suspicious copy of unauthorized IP. Without an effective copy detection method, all the previous efforts in watermarking and fingerprinting are in vain. Although there are some progress in copy detection, we argue that to assure fast detection, the watermark/fingerprint for copy detection methods are required, which hide the marks behind certain parts of the problem with rather unique structure that are difficult to be altered.

This book contains the mathematical foundations for the developed IP protection paradigm, detailed pseudocode and descriptions of its many techniques, numerous examples and experimental validation on well-known benchmarks, and clear explanations and comparisons of the many protection methods that can be applied for the protection of VLSI design IPs from FPGA design to standard-cell placement, from high-level synthesis solutions to gate-level netlist place-and-rout, and from advanced CAD tools to physical design algorithms.

We conclude that the essence of this IP protection technique is constraint manipulation. Although we restrict our discussion to VSLI design IP protection, constraint manipulation is a method that has a much broader range of applications. For example, in the field of applied cryptography and computational security, one can build constraint-based protocols for privacy protection, access denial, and so on; another applicable area is applied optimization algorithm, where one can implement software to tackle hard optimization problems (such as graph coloring and traveling salesman problems). More specific, in these problems, the search for better solutions is usually expensive and sometimes the same solution may be visited more than once from different search paths. Introducing additional constraints to the search process will make the search more efficient and effective.

Appendix A
Intellectual Property Protection: Schemes, Alternatives and Discussions

Issued by Intellectual Property Protection Development Working Group
Released August, 2000

Revision 08 Jan. 2001

This appendix is part of the VSI Alliance™ White Paper (IPPWP1 1.1) issued by Intellectual Property Protection Development Working Group. We gratefully thank VSI Alliance and Mr. Ian R. Mackintosh, Chair of the IPP Development of Working Group and the author of the white paper, for granting us the permission to include this white paper in the book. The whole document is available at the Alliance website www.vsi.org.

VSIA IP PROTECTION DWG

By late 1999, VSI Alliance™ (VSIA) had established eight Development Working Groups (DWG's) each strongly supporting the VSIA vision:

"To dramatically accelerate system chip development by specifying open standards that facilitate the mix and match of virtual components (VCs) from multiple sources."

The Intellectual Property Protection (IPP) DWG was created in 1997 to address the issue of protection of virtual components (VCs). The goals of this DWG were to:

- Enable IP Providers to protect their VCs against unauthorized use
- Protect all types of Design Data used to produce and deliver VCs
- Detect use of VCs
- Trace use of VCs

SCOPE

Various solutions exist for protection of virtual components (VCs), but not all are equally applicable to each type of VC. Trade-offs exist between the value (perceived or real) of the VC, difficulty of implementation of the protection scheme, and the resulting usability of the protected VC by both the integrator and the end user. This paper briefly discusses and introduces known technologies and mechanisms that support the broad spectrum of VC types, sources of VCs, and business requirements for VC users and providers.

The scope of this paper is to identify open, interoperable, standards-based solutions (or guidelines and information where standards are not practical) for VC protection which balance the level of security with customer usability of VCs, while fostering design reuse from creation through to the effective use of VCs. In this context, "VC" includes products, technology and software that may be protected through patents, copyrights or trade secrets. The trade-offs discussed can be used in selecting appropriate protection mechanisms for hard, firm and soft VCs.

The broad target audience for this paper includes VC providers, VC users (system designers or integrators), EDA vendors, and semiconductor vendors who utilize virtual components in standard product FPGA, CPLD, ASIC, or SoC market segments. Various protection, detection, and tracking mechanisms that can be employed with VCs and that are licensable to another party are discussed. This paper is concerned with protection of VCs and not with protection of design programs (EDA tools) used in processing them through a design flow.

INTRODUCTION

The general infringement of all types of intellectual property (IP) in the United States has become a major problem. At the 1998 annual RSA Conference, it was estimated that the cost of IP infringement approaches $1 billion per day. This problem has received so much attention that the FBI launched Operation Counter Copy to address it. Today, the FBI estimates that 80 percent of all infringements of electronic designs can be traced to sources from within the company that developed the IP. The other 20 percent occur at external points of vulnerability, caused by the ease with which end-products can be reverse engineered, copied or simply stolen.

In the area of electronic design, there are an estimated 100 reverse engineering shops in the US; approximately 70 percent of these are funded by government(s), and many of the techniques developed are leaked, or even published, to the industry. The American Society for Industrial Secrets estimates that in the US alone, trade secret theft is in excess of $2 billion per month.

Although the protection of VCs is rapidly becoming a major concern within the VC and Electronics Industry as a whole, the overall awareness of the issue remains low. As the electronics industry shifts to a design-for-reuse methodology, virtual component trading is expanding, and the potential for infringement (intentional or unintentional) is growing in proportion. Unfortunately, awareness of the liabilities may only be achieved in the aftermath of a highly visible, industry scare.

How, then, can virtual components be protected? Unfortunately, potential infringers have the upper hand today, with so few IP protection programs in place. In truth, it is almost impossible to guarantee protection of a VC in all of its uses, data forms, and exposures during use. However, it is realistic to define and apply adequate mechanisms and precautions such that the costs for infringers exceeds the value of success and the cost of the protection afforded to VC owners is consistent with the risk and value of loss.

An early example of an IP Protection scheme was to have EDA tools create and operate on an encrypted form of the source code of a virtual component. However, encryption supported by EDA tools has inherent flaws (see the section on Protection Mechanisms):

- EDA tool vendors do not license encryption algorithms to others.
- The author of the VC must trust and rely upon the EDA vendor's security, since the EDA vendor retains decryption capability.
- All EDA tools have back-door access to the encrypted data in order to determine if problems encountered are due to a bug in the tool or the VC.

It is essential that practical solutions support both customer use and supplier distribution models in the form of recommended guidelines, practices, standards and implementation plans.

OVERVIEW: Security Schemes

There are three approaches to the problem of securing a VC. Using the deterrent approach, the VC owner may deter the infringer from contemplating the theft of the VC by using proper legal means. With the protection approach, the owner tries to prevent unauthorized use of the VC. And, using the detection approach, the owner detects and traces both legal and illegal use of the VC, so that a proper course of action can be taken.

Deterrents provide external communication of legal protection in an attempt to deter an illegal act from occurring. They do not provide any physical protection. Types of deterrents available include:

- Patents
- Copyrights
- Trade Secrets
- Contracts and Lawsuits

Protection involves taking active steps to try to prevent the unauthorized uses of VC's from occurring. Protection mechanisms include such tangibles as:

- Licensing Agreements
- Encryption

Detection involves the ability to determine that an unauthorized use has occurred and then, tracing the source of the theft. Detection and traceability methods that are becoming available include:

- Foundry IP Tracking or Tagging (see VSIA's, Virtual Component Identification: Physical Tagging Standard)
- Digital Signatures, such as, Digital Fingerprinting and Digital Watermarking
- Noise Fingerprinting

Ideally, a trace would be created every time a VC is used in any form during design, implementation or fabrication. Information would be logged and carried along with other data including tool use, user identification, time, date, etc. For designers (users of VC's), assembly of multiple VCs requires that auditing be made hierarchical. Such an ideal system would uncover theft and provide notification back to the VC Provider.

Security Schemes appropriate for a VC are determined by the specific application point of the VC during its life-cycle. A VC evolves through phases of development, licensing, use, and sales of an end product; and, discovery of an infringed property can occur anywhere in that evolution process. At specific points of this life-cycle, different security schemes will need to be implemented. An example of these schemes and the life-cycle phase of a VC is shown below.

DETERRENTS

Traditional deterrent protection mechanisms are patents, copyrights, trademarks and trade secrets. The primary goal of patents and copyrights is to encourage commercialization and give exclusive rights to the originator for a specific period of time. These methods provide varying degrees of protection, especially in the international community.

A developer needs to understand the regulations and principles behind the methods of providing protection to VC designs, both for the protection of the developers own designs and for the protection of other developer's virtual components. A detailed search and analysis of patents, copyrights and trademarks should be conducted prior to initiating any VC developments, to establish any potential infringements of other intellectual property rights, and to aid in determining the worth of the developer's proposed VC design.

		D	L	I	M	U	A	ID
	Patents	X		X				
	Copyrights	X		X				
DETERRENTS	Trade Secrets	X		X				
	Contracts		X					
	Lawsuits							X
PROTECTION	Licensing Agreement		X			X	X	
	Encryption	X						
	Tracking				X			X
	Tagging	X		X				X
DETECTION	Digital Fingerprinting	X		X				X
	Digital Watermarking	X		X				X
	Noise Fingerprinting	X		X				X

Table A.1. Example Security Schemes Applicable During VC Life-Cycle: D = Development, L = Licensing, I = VC Integration, M = Manufacture, U = End Component Use, A = End Application, ID = Infringement Discovery.

Patents

It is important to note that patents are only recognized in the specific country where the patent was filed. Typically, a US patent costs $10K-$30K (including prosecution and lifetime maintenance fees) and is applicable (active) for up to 20 years. An international patent costs approximately $50K-$100K (including prosecution, translations and annual annuities over the life of the patent) with varying duration of protection.

A patent also requires extensive documentation. The author must prove novelty and utility and give complete directions for implementing the invention. Once a patent is issued, it is fully disclosed to the public. If an international application for patent protection in other countries under their laws is not submitted, these patents will be protected only in the country of application.

Copyrights

Copyrights were originally designed to protect literature, music and dramatic works. They only prohibit copying expressions of an idea, not the idea itself (as a patent does). Therefore, it is easier to get a copyright than a patent. Copyrights have a much longer period of protection (50 years beyond the life of the author), and they are recognized internationally. However, international laws make them difficult to enforce. With respect to semiconductor designs, copyrights have only limited use. They are generally applied only to the die or masks to prevent exact copies.

Trade Secrets

A trade secret law has a broader scope of coverage than patents and copyrights. However, the author must take deliberate steps to protect and secure the information in order to be covered by trade secret laws. The author must also derive economic benefit from the secret information. Typically, trade secrets are created and owned by companies, rather than individuals. Trade secrets are kept by the originator to maintain exclusive rights. A prime example is the recipe for Coca-Cola. It is not only a trade secret, but no one person knows the whole recipe.

To receive protection under trade secret laws, a company must restrict access to information being held as a trade secret. If the information must leave the premises, intent must be shown to protect and control the data. In regard to contracts with other companies, trade secrets must be described in detail, the rights being granted must be well-defined, and the information must be declared to be held as a trade secret. Access to trade secret information must be carefully and consistently documented. As noted previously, the major hole in security is from within companies. So, it is imperative that the employees sign employment contracts stipulating the company policy on trade secrets. If the information becomes public, trade secret law cannot be used as protection.

Governing Law

It is also very important to understand the nature and scope of the jurisdiction that provides the various types of protection, since laws are made and adjudicated by different government organizations. The following chart is used to illustrate the diversity between governing bodies and is not to be interpreted as a comprehensive summary of the worldwide laws that govern the protection of Intellectual Property.

GOVERNMENT	COPYRIGHT	TRADEMARK	PATENT	TRADE SECRET
US-Federal	Yes/50-100 Years	Yes/Permanent	Yes/17-20 Years	No-Guidelines
US-State	No	Yes (Varies)	No	No-Guidelines
Foreign	Yes/No	Yes/Some	Yes/Some	No

It should be noted that Intellectual Property rights in all cases except those involving trade secrets are affirmative rights, which means that the burden is upon the owner to initiate action against the infringer in cases of alleged infringement of a patent, copyright or trademark. On the other hand, trade secrets are often treated in the same manner as tangible property rights, which means that the authorities may take action against the accused party under criminal law, if the owner reports a theft or loss. The burden of pursuing affirmative rights rests with the owner of the Intellectual Property.

PROTECTION MECHANISMS

For highly proprietary VCs of great value, loss of control of EDA design data could result in large financial losses. So, it is important to protect these VCs with a high degree of security, such as that provided by encryption. At the same time, it is prudent to provide customers with a means of evaluating potential VC purchases, prior to the actual purchase.

Encryption

Encryption provides a means of giving potential customers access to an executable version of a VC without specific access to the source code. This mechanism allows recipients to try the VC, integrate it, and process through the various EDA tools in the flow towards silicon manufacturing without specifically disclosing the structure of the VC to the customer.

The problem is that not all EDA tool vendors provide tools supporting encryption; encryption is often proprietary, and there exists built-in "back-doors" to EDA tools that could permit a user to gain access to the unencrypted source code. As more EDA vendors establish their VC protection philosophy and strategy, the power of encryption could become more available and viable in supporting VCs, despite problems of customer willingness to pay for such capabilities.

Not all encryption schemes are optimal and any scheme employed should pass minimum tests of usability. For example, the public domain Pretty Good Protection (PGP) encryption scheme has been considered as a low-cost, open method to protect the distribution and exchange of VCs.

However, there is currently insufficient infrastructure and control over the use of keys, which diminishes the value and potential in this application.

Hardware Protection

A powerful means of directly protecting EDA design data of a VC is simply not to release the design data, except in more indirect forms:

a) in the form of GDS II tapes (under foundry control) to make masks for the complete chip, or

b) in the form of a programmable device such as an FPGA (see section on Silicon Security), for use in a hardware or emulation platform.

Neither of these forms permits access to complete design views, and both of these methods increase the level of difficulty in gaining access to source information defining the VC.

Chemical Protection

Passivation technology was developed to protect the actual silicon die from the reverse engineering process. Much of this work was carried out by the military and involves the creation of inert passivation applied to the silicon as part of the normal manufacturing process. The passivation acts in its usual, protective fashion unless its surface is scratched and exposed to the atmosphere. When this happens, the passivation becomes reactive and damages the exposed silicon, preventing reverse engineering.

DETECTION SCHEMES

Various mechanisms exist to allow the identification of ownership of a VC. These schemes afford differing levels of security; some are deeply and undetectably buried in a design and others are openly displayed, easy to observe, and used as a simple means of tracing a VC. The most well-known schemes are described below.

Tagging and Tracking

Tagging and tracking are simply attaching tags or labels to VCs for tracing these elements (generally in the manufacturing phase) and enabling honest people to keep appropriate records and conduct their business efficiently and safely.

An example of such a scheme is the VSIA's, IPP DWG sponsored "Virtual Component Identification: Physical Tagging Standard", available to both VSIA members and non-members. This technique simply creates a GDSII label for any VCs grouped on an IC design. This label (or "tag") contains information on title, ownership, origination date, number of occurrences, etc. and permits an entity, such as a silicon foundry, to record uses, recognize ownership and administrate events and royalty payments.

Alternative tagging technologies are emerging, such as that from SIIDTECH (Portland, Oregon), which permits the unique and repeatable creation of digital ID's for individual silicon die. This patented technology offers a drop-in GDSII cell for the silicon die that features single pad readout of a non-volatile signature ID. It is technology for the physical silicon level of abstraction, equally useful to both foundries wishing to record unique identifiers for individual wafers, or to markets demanding individual identification and tracking of silicon die.

It is likely that in the future, infrastructure will emerge whereby an independent body will carry records of IP ownership, labeling, tagging and even digital signatures. Such an enterprise would be similar to that already existing in the music industry, where royalties for the use of music are collected and distributed to both users and owners of that music, who are due royalties.

Digital Signatures
A VC has a digital signature or fingerprint, which is a characteristic of the VC that acts as a virtually unique and exclusive identifier. More accurately, a digital signature is a finite, possibly hierarchical sequence of symbols drawn from a finite alphabet.

The fingerprint is generally the indigenous characteristic of a VC, whereas a signature can be the representation of that fingerprint, whether it is indigenous, or artificially inserted in the VC for purposes of identification or tracking.

Digital Fingerprinting
Digital fingerprinting is sometimes called passive watermarking. Here the recording and extraction of the unique digital signature utilizes inherent, pre-existing characteristics or attributes of a VC. The signature is a representation of the unique features and overall structure of the VC. Essentially, the mechanism is like a lossy compression scheme, where a complex and possibly hierarchical VC is characterized into a single digital signature.

The benefits of the scheme include avoidance of tampering with or changing of the VC, the use of standard design flows, and speed of implementation without performance hits. Fingerprints do NOT lend themselves to reverse engineering of the VC and are very suitable to be collected in databases (a la FBI fingerprinting). Such unique identifiers could find application as keys in encryption schemes.

Limitations include the fact that a fingerprint does not carry with it such useful information as the owner, VC name, etc. and so has some weakness relative to a simple tagging mechanism. A simple revision of a VC establishes a new fingerprint.

It is possible to record the digital fingerprint of a VC at most levels of abstraction in the design hierarchy. The VSIA IPP DWG plans to publish further work on digital fingerprinting during the year 2000.

Digital Watermarking
Digital watermarking is an indirect protection scheme in that it provides a deterrent to infringers by offering the ability to demonstrate ownership of a VC to its originator. The process of active watermarking consists of the implantation of a digital signature into a VC at a particular level of design abstraction, while utilizing the intrinsic features and structure of that level.

Watermarking is a hot area for research both within industrial and academic circles. Promising recent work suggests that efficient tools and methods are emerging to make the cost of both implementation and detection of watermarks economically feasible in the not-too-distant future.

Hierarchical watermarking is a scheme that targets more than one abstraction level for the same VC. Watermarks have been demonstrated at the highest level of algorithmic abstraction and propagated down to the physical level. An example might be the encoding of a digital pattern of "1's" and "0's" in the pass band of a complex filter, that can be observed (for example) in the frequency spectrum of that filter in the physical domain. A further example would be the encoding of an extractable pattern in a piece of logic that utilizes unused state transitions to implement the watermark; undetectable to all but those most intimately familiar with the VC.

The key challenges in this area are to develop tools and methods that are extremely difficult to defeat, have low cost/performance penalties, do not impede the native operation of the system, and are intuitively acceptable as proof of ownership in a court of law.

Additionally useful characteristics of watermarks include their holographic nature. It is possible to employ a watermark (a digital signature) broadly across a whole design, within single or multiple VC's, or even inside small functional areas. This practice means that small or even large portions of a design cannot be copied without the risk of traceable watermarks remaining undisturbed and verifiable within their new and illegal application.

Noise Fingerprinting

Noise fingerprinting is another passive scheme for identifying digital circuits. Here the switching activity within a circuit causes a unique noise signature into the silicon substrate, with a resultant spectrum for the signature being determined by process variations, input sequences, and circuit implementation specifics.

Particular input stimuli can be generated for a VC or design and the resultant noise characteristics are observed through substrate pads, pick-ups, or supply lines. These fairly exotic concepts can be implemented without requiring many of the expensive forensic technologies often customary when checking for unauthorized use of VCs and whole designs within fabricated chips.

SILICON SECURITY

The following discussions review some of the most popular forms of silicon implementation for VCs. It is generally possible to reverse engineer and extract intellectual property from each type of silicon technology - the issue is the degree of difficulty for each type.

Extracting a whole VC from silicon can be more difficult than reverse engineering the entire functionality of the silicon die. This is because a VC realized in silicon can be physically merged with other functions or, (for example) be just an embedded part of a larger bit-stream. So, reverse engineering an entire silicon die or function is one thing, but it requires different and more VC-specific knowledge to extract a particular VC. The following are some silicon technologies explained in more detail to illustrate how this applies:

Programmable SRAM Devices.

Many designs today are utilizing programmable logic to speed their time to market. Programmable devices based on SRAM are volatile; meaning the configuration data is lost each time the device loses power (whether intentionally or because of power interruption.) SRAM-based devices typically store the configuration information in an external location, such as a serial PROM or microprocessor code space, which is downloaded each time the device is powered-up.

There are two techniques used to copy SRAM-based programmable designs: either duplicate the PROM, or duplicate the configuration bit-stream and program the other devices. Either approach can be accomplished quickly and easily. While this technique would allow the illegal copies of a complete SRAM FPGA, a specific IP implemented in the design is not compromised. Extracting Intellectual Property requires an additional and more sophisticated technique. Not only is the capture of the download configuration information needed, but so is the internal logic structure of the SRAM-based FPGA itself, to determine the function performed as a result of the programming. Since most SRAM-based programmable logic has a regular structure, this can be determined for a given architecture, with appropriate investment in reverse engineering. Internal logic structures are proprietary and are unpublished, and while it may be cost effective to reverse engineer a 3000 - 5000 gate design, it is a daunting task to extract Intellectual Property from a flat netlist of 1-2 million gates. An engineering team might often be better off creating their own block diagrams and developing their own VC implementation.

Hard-Mask ICs

It is popularly believed that the most difficult programmable device to reverse engineer is a hard mask IC. However, due to the need for failure analysis tools, the industry has developed many sophisticated techniques to reverse engineer a hard mask IC. One technique is to selectively strip off one layer at a time, photographing the layers as they are exposed. These photographs are then overlayed, and the interconnect and transistors are extracted from the design. (See the section on, Chemical Protection, which would prevent this approach from being taken.)

An experiment was performed utilizing this technique, which showed that it took two weeks to reverse engineer and capture an entire 386 processor. This experiment showed that if a complete chip is reverse engineered, a copy can be made. A more difficult task is to extract individual VCs so that they can be independently used in a different design. So, while hard mask integrated circuits are more secure than SRAM or flash-based technologies, extraction and use of a particular VC netlist comprised of 10K's to millions of gates, (when logical functions may be physically merged), may require comparable expertise to creating the VC from scratch.

Antifuse Programmable Devices

Once programmed, an antifuse is inherently non-volatile, which allows the device to retain its configuration indefinitely without external means-batteries, PROM or microprocessor code space. Antifuses do not have any residual electric or magnetic fields to detect, nor is there anything visual that can be seen from the top or bottom of the die to determine the programmable state of the antifuse device locations.

The only successful attempts at locating programmed antifuses has been using a Transmission Electron Microscope (TEM). This is a destructive sample technique that costs approximately $1,000 for a single TEM sample, today. With approximately 500,000 antifuse sites on a typical antifuse part, it would cost at least $500 million to capture a complete design. Furthermore, to capture the design, 20,000 programmed antifuses would have to be identified exactly to copy or reverse engineer a single sub-10K gate design. A limitation of antifuse technology is the relatively low gate count of 50-100K gates. Even though there are no known efficient techniques to reverse engineer antifuse technology, some antifuse providers have already provided the ability to incrementally change FPGA die areas in such a way as to permit the insertion of digital signatures or keys on a chip-by-chip basis.

So, be aware that when a native implementation technology (such as SRAM, hard-mask, antifuse, flash, etc.) is selected, there is an inherent ease/difficulty in extracting both the entire design and also in extracting a specific portion of that design (such as a single VC).

CLOSING DISCUSSION

It is up to each developer, owner or licensee of virtual components to determine the type and amount of protection that will be employed for each VC in their possession. The party needs to have an assessment made of the actual and strategic value of each VC design in order to determine the type of protection or control dictated for the VC. How important is the design (VC) to the company and what is the cost of the potential loss of control of the VC? The owner needs to understand the regulations and principles of the methods providing protection to VC designs. Where will the user of the VC integrate, fabricate and sell the chips that are generated using the developer's VC?

It is important to understand the nature and type of protection that should be afforded to the VC, since legal organizations that operate under different governments may be called upon to adjudicate the improper use of the VC. Therefore, very significant factors in the licensing of virtual components are not only a complete understanding of their use and application, but also the development of a high-level of mutual trust with the licensee.

Based on a careful assessment of the above, the owner of VCs must then decide which forms of protection and care provide the best security and level of risk for releasing the VC to a third (and fourth, etc.) party, given the value of the sale/trade. The tradeoffs used in making this type of decision are often unique and may be specific to each developer, user, and also to each virtual component developed and licensed.

Owners should generate a matrix for each virtual component that documents and analyses the following categories of exposure, in order to assess the type of protection that is appropriate for each element of a virtual component.

The chart below shows an example of how one might evaluate, and afford protection to, a given virtual component. The value statement is that of the particular element of the virtual component to the owner. There are no fixed rules that can be used in making this type of assessment, because considerations are all relative to the owner's business, their personal and technical judgements, and the projected effect upon current and future revenue and profit potential for the company. It is likely that over time, some of the actions relative to the considerations will change and so, any matrix such as this will need to be updated and maintained.

Item	Protection	Patent	Protection Type	Item Value	Protection Cost
System Evaluation Model	No	No	LA	$	$
Test Bench	No	No	LA	$	$
Behavioral Model	No	No	LA or DF	$	$
Bus Model	Yes	No	LA or DF	$	$
Schematic Diagram	Yes	Possible	LA	$	$
RTL Source	Yes	Possible	LA,E,DW	$	$
Physical Netlist	Yes	Possible	F	$	$
GDS II Tape	Yes	Possible	LA,E,DW	$	$
	Total Sum Cost of Protection Schemes				$
			Company Value of VC	$	

Table A.2. Example VC Protection Scheme Summary: LA = Legal Agreement, DF= Digital Fingerprint, DW= Digital Watermark, E= Encryption, F= Antifuse FPGA.

In addition to the decision on the investment of protection schemes for a given VC, such judgements should be preceded by understanding such issues, as:

a) Where will these various levels of abstraction reside?

b) Who will and should have access to this data?

c) How will the environment be secured?

d) How will data in transit be protected?

e) How are tools manipulating the data secured?

Not every company can practically afford to guard against all potential liabilities and implement exhaustive protection schemes. However, it is prudent that every company should understand the scope of its liabilities and be proactive in the selection of their intellectual property protection schemes.

In a closing observation, one would consider it imprudent, for example, if the head of a household did not carry insurance for the home, an event of death, or loss of the family car. Why then would responsible executives and managers not protect their investors by thoughtfully securing the intellectual property of their company?

References

[1] A. Adelsbach, B. Pfitzmann, and A. Sadeghi. "Proving Ownership of Digital Content". *The 3rd International Information Hiding Workshop*, pp. 126-141, September 1999.

[2] A.V. Aho, "Algorithm for Finging Patterns in Strings," *Handbook of Theoretical Computer Science*, 1990.

[3] C. J. Alpert, "Partitioning Benchmarks for the VLSI CAD Community", Web page, http://vlsicad.cs.ucla.edu/~cheese/benchmarks.html

[4] C. J. Alpert, "The ISPD-98 Circuit Benchmark Suite", *Proc. ACM/IEEE International Symposium on Physical Design*, April 98, pp. 80-85. See errata at http://vlsicad.cs.ucla.edu/~cheese/errata.html

[5] C. Ajluni, "Redefining EDA in the New Age of Intellectual Property," *Electronic Design*, Vol. 46, No. 1, pp. 64-76, January 1998.

[6] D. Aucsmith. "Tamper Resistant Software: an Implementation", *1st Information Hiding Workshop*, Lecture Notes in Computer Science, Vol. 1174, pp. 317-334, Springer-Verlag, 1996.

[7] B.B. Ames. "Shortening the Design Cycle: You Want It When?" *Design News*, http://www.manufacturing.net/magazine/dn/archives/2000/dn0221.00/feature2.html, February 2000.

[8] R. Anderson and M Kuhn, "Tamper Resistance – A Cautionary Note", *USENIX Workshop on Electronic Commerce*, pp. 1-11, November 1996.

[9] T. Aura and D. Gollman, "Software Licence Management with Smart Cards", *Proceedings of the USENIX Workshops on Smartcard Technology*, pp. 75-85, May 1999.

[10] B. S. Baker and U. Manber. "Deducing similarities in Java sources from byte-codes", *USENIX Technical Conference*, pp. 179-190, 1998.

[11] R. Bayardo Jr. and R. Schrag, "Using CSP look-back techniques to solve exceptionally hard SAT instances", *Principles and Practice of Constraint Programming*, pp. 46-60, 1996.

[12] R.Bayardo Jr. and R. Schrag. "Using CSP Look-Back Techniques to Solve Real-World SAT Instances," *Proceedings of the National Conference on Artificial Intelligence (AAAI'97)*, 1997.

[13] P. Benassi, "TRUSTe: An Online Privacy Seal Program," *Communications of ACM*, Vol.42, No.2, pp. 56-59, Febuary 1999.

[14] W. Bender, D. Gruhl, N. Morimoto, and A. Lu. "Tehcniques for Data Hiding," *IBM Systems Journal*, Vol. 35, No. 3&4, pp. 313-336, 1996.

[15] G. Benson, " An Algorithm for Finding Tandem Repeats of Unspecified Pattern Size ," *Proc. RECOMB98 Second Annual International Conference on Compu-tational Molecular Biology*(S. Istrail, P. Pevzner, M. Waterman, eds.), 1998, p. 20-29.

[16] H.Berghel and L.O'Gorman. "Protecting ownership rights through digital watermarking," *IEEE computer,* Vol. 29, No. 7, pp. 101-103, July 1996.

[17] D. Bertsekas and R. Gallager, *Data Networks*, Prentice-Hall, 1987.

[18] I. Biehl, and B. Meyer. "Protocols for Collusion-Secure Asymmetric Fingerprinting," *STACS'97, Proceedings of 14th Annual Symposium on Theoretical Aspect of Computer Science*, Reischuk, and Morvan (Eds.), Springer-Verlag pp. 399-412 1997.

[19] B.Bollobás. "Random Graphs," Academic Press, London, 1985.

[20] D. Boneh, and J. Shaw. "Collusion-Secure Fingerprinting for Digital Data," *Advances in Cryptology - CRYPTO'95, Proceedings of 15th annual International Cryptology Conference, Coppersmith (Ed.)*, Springer-Verlag, pp. 452-465 1995.

[21] L.Boney, A.H.Tewfik, and K.N.Hamdy. "Digital watermark for audio signals," *International Conference on Multimedia Computing and Systems*, pp. 473-480, 1996.

[22] R. S. Boyer and J. S. Moore, "A Fast String Searching Algorithm ," *Communi-cations of the ACM*20(10), 1977, pp. 762-772.

[23] D. Brelaz, "New methods to color the vertices of a graph", *Communications of the ACM*, Vol.22, No.4, pp. 251-256, 1979.

[24] F. Brglez and H. Lavana. "A Universal Client for Distributed Networked Design and Computing", *38th ACM/IEEE Design Automation Conference Proceedings*, pp. 401-406, June 2001.

[25] S. Brin, J. Davis, and H. Garcia-Molina, "Copy detection mechanisms for digital documents." *SIGMO Record*, Vol. 24, No. 2, pp. 398-409, 1995.

[26] R.E. Bryant, "Binary decision diagrams and beyond: enabling technologies for formal verification", *ICCAD*, pp. 236-243, 1995.

[27] A.E. Caldwell, H. Choi, A.B. Kahng, S. Mantik, M. Potkonjak, G. Qu, and J.L. Wong. "Effective Iterative Techniques for Fingerprinting Design IP," *36th ACM/IEEE Design Automation Conference Proceedings*, pp. 843-848, June 1999.

[28] F.L. Chan, M.D. Spiller, and A.R. Newton. "WELD - An Environment for Web-Based Electronic Design", *35th ACM/IEEE Design Automation Conference Proceedings*, pp. 146-151, June 1998.

[29] M.T.Chao, and J.Franco. "Probabilistic Analysis of Two Heuristics for the 3-Satisfiability Problem," *SIAM Journal of Computing*, Vol.15, No.4 pp. 1106-1118, 1986.

[30] E. Charbon. "Hierarchical Watermarking in IC Design," *IEEE 1998 Custom Integrated Circuits Conference*, pp. 295-298, 1998.

[31] E. Charbon and I. Torunoglu, "Watermarking layout topologies", *ASPDAC*, pp. 213-216, 1999.

[32] P. Cheeseman, B. Kanefsky, and W.M. Taylor. "Where the Really Hard Problems Are," *Twelveth International Joint Conference on Artificial Intelligence*, pp. 331-337, 1991.

[33] P. Chen and K. Keutzer. "Towards True Croostalk Noise Analysis," *IEEE/ACM International Conference on Computer Aided Design*, pp. 132-137, November 1999.

[34] K.-W. Chiang, S. Nahar and C.-Y. Lo, " Time-Efficient VLSI Artwork Analysis Algorithms in GOALIE2 ," *IEEE Transactions on Computer-Aided Design of Integrated Circuits and Systems*8(6), 1989, pp. 640-648.

[35] B. Chor, A. Fiat, and M. Naor. "Tracing Traitors," *Advances in Cryptology - CRYPTO'94, Proceedings of 14th annual International Cryptology Conference. Desmedt (Ed.)*, Springer-Verlag, pp. 257-270, 1994.

[36] B. Cmelik and D. Keppel, " Shade: a fast instruction-set simulator for execution profiling," *SIGMETRICS Conference on Measurement and Modeling of Com-puter Systems*22(1), 1994, pp.128-37.

[37] A. Cohen, "Spies among Us," *Time Digital*, pp. 32-39, July 2000.

[38] C. Collberg, C. Thomborson, and D. Low, "A Taxonomy of Obfuscating Transformations", *Technical Report #148*, Department of Computer Science, University of Auckland. July 1997.

[39] C. Collberg, C. Thomborson, and D. Low, "Manufacturing Cheap, Resilient, and Stealthy Opaque Constructs," *Symposium on Principles of Programming Languages,*1998, pp. 184-196,

[40] C. Collberg and C. Thomborson. "Software Watermarking: Models and Dynamic Embeddings", *ACM Symposium on Principles of Programming Languages*, January, 1999.

[41] M. R. Corazao, M. A. Khalaf, L.M. Guerra, M. Potkonjak and others, " Per-formance Optimization Using Template Mapping for Datapath-Intensive High-Level Synthesis ," *IEEE Transactions on Computer-Aided Design of Integrated Circuits and Systems*15(8), 1996, pp. 877-888.

[42] O. Coudert, "Exact Coloring of real-life graphs is easy", *34th Design Automation Conference*, pp. 121-126, June 1997.

[43] I.J.Cox, J.Kilian, T.Leighton, and T.Shamoon. "A secure,imperceptible yet perceptually salient, spread spectrum watermark for multimedia," *Southcon*, pp. 192-197, 1996.

[44] S. Craver. "Zero Knowledge Watermark Detection". *The 3rd International Information Hiding Workshop*, pp. 102-115, September 1999.

[45] S. Craver, N. Memon, B.L. Yeo, M.M. Yeung, "Can invisible watermarks resolve rightful ownerships?" *Technical report, IBM Research Technical Report RC 20509*, 1996.

[46] J.M. Crawford, "Solving Satisfiability Problems Using a Combination of Systematic and Local Search", *Second DIMACS Challenge*, 1993.

[47] M. Dalpasso, A. Bogliolo, and L. Benini, "Virtual Simulation of Distributed IP-Based Designs", *36th ACM/IEEE Design Automation Conference Proceedings*, pp. 50-55, June 1999.

[48] M. DeGroot, *Probability and Statistics*, Addison-Wesley, Reading, 1989.

[49] J. Domingo-Ferrer, "Anonymous Fingerprinting of Electronic Information with Automatic Identification of Redistributers," *Electronics Letters*, Vol.34, No. 13, pp. 1303-1304, 1998.

[50] S. Dutt and W. Deng, "VLSI Circuit Partitioning by Cluster-Removal Using Iterative Improvement Techniques", *Proc. IEEE International Conference on Computer-Aided Design*, 1996, pp. 194-200.

[51] L. Entrena and K.-T. Cheng. "Sequential Logic Optimization by Redunancy Addition and Removal." *IEEE/ACM International Conferenc eon Computer AIded Design*, pp. 310-315, November 1993.

[52] M. Fang, N. Shivakumar, H. Garcia-Molina, R. Motwani and J. Ullman, "Computing Iceberg Queries Efficiently ," *Proc. International Conference on Very Large Databases,*New York, August 1998.

[53] C. M. Fiduccia and R. M. Mattheyses, "A Linear Time Heuristic for Improving Network Partitions", *Proc. ACM/IEEE Design Automation Conference*, 1982, pp. 175-181.

[54] A. Fin and F. Fummi. "A Web-CAD Methodolgoy for IP-Core Analysis and Simulation", *37th ACM/IEEE Design Automation Conference Proceedings*, pp. 597-600, June 2000.

[55] C. Fleurent and J. A. Ferland. "Genetic and hybrid algorithms for graph coloring." *Annals of Operations Research,* Vol. 63, pp.437-461, 10067.

[56] D. A. Forsyth and M. M. Fleck, "Finding People and Animals by Guided As-sembly ," *Proc. International Conference on Image Processing,*1997, vol. 3 pp. 5-8.

[57] J. Franco, and M. Paull. "Probabilistic analysis of the Davis Putnam procedure for solving the satisfiability problem," *Discrete Applied Mathematics*, Vol. 5, pp. 77-87, 1983.

[58] J. Franco, and Y.C. Ho. "Probabilistic Performance of A Heuristic for the Satisfiability Problem," *Discrete Applied Mathematics*, Vol. 22, pp. 35-51, 1988.

[59] J. Franco. "Elimination of infrequent variables improves average case performance of satisfiability," *SIAM Journal on Computing* Vol. 20, No. 6, pp. 1119-1127, December 1991.

[60] E. Gabber, P.B. Gibbons, D.M. Kristol, Y. Matias, and A. Mayer, "Consistent, Yet Anonymous, Web Access with LPWA," *Communications of ACM*, Vol.42, No.2, pp. 42-47, Febuary 1999.

[61] M.R. Garey and D.S. Johnson. "Computer and Intractability: A Guide to the Theory of NP-Completeness," W.H. Freeman and Company, New York, NY, 1979.

[62] M.X. Goemans and D.P. Williamson, "Improved approximation algorithms for maximum cut and satisfiability problems using semidefinite programming", *Journal of the ACM*, Vol.42, No.6, pp. 1115-1145, 1995.

[63] A. Goldberg. "On the complexity of the satisfiability problem," *Courant Comp. Sci. Rep.*, No. 16, New York University, New York, 1979.

[64] A. Goldberg, P.W. Purdom Jr., and C.A. Brown. "Average time analysis of simplified Davis-Putnam procedure," *Information Process Letters*, Vol. 15, pp. 72-75, 1982.

[65] D. Goldschlag, M. Reed, and P. Syverson, "Onion Routing for Anonymous and Private Internet Connections," *Communications of ACM*, Vol.42, No.2, pp. 39-41, Febuary 1999.

[66] S. Grier, " A Tool that Detects Plagiarism in PASCAL Programs," (12th SIGCSE Technical Symposium on Computer Science Education, St. Louis, Feb. 1981), *SIGCSE Bulletin*13(1), 1981, pp. 15-20.

[67] D. Grover. "Forensic copyright protection." *Computer Law and Security Report*, Vol. 14, No. 2, pp. 121-122, 1998.

[68] M. Haertel, et al, "The GNU diff program," Available by anonymous FTP from prep.ai.mit.edu, 1999.

[69] M.M. Halldorsson, "A still better performance guarantee for approximate graph coloring", *Information Processing Letters*, Vol. 45, No. 1, pp. 19-23, 1995.

[70] F. Hartung, P. Eisert, and B. Girod, "Digital Watermarking of MPEG-4 Facial Animation Parameters," *Computer Graphics*, Vol. 22, No. 3, pp. 425-435, 1998.

[71] F. Hartung, and B. Girod, "Digital watermarking of raw and compressed video," *In Proceedings of the SPIE-The Internation Society for Optical Engineering*, Vol. 2952, pp. 205-213, 1996.

[72] F. Hartung and B. Girod. "Fast Public-Key Watermarking of Compressed Video". *IEEE International Conference on Image Processing*, pp. 528-531, October 1997.

[73] K. Hines and G. Borriello. "A Geographically Distributed Framework for Embedded System Design and Validation", *35th ACM/IEEE Design Automation Conference Proceedings*, pp. 140-145, June 1998.

[74] I. Hong, M. Potkonjak. "Technique for Intellectual Property Protection of DSP designs," *ICASSP98 International Conference on Acoustic, Speech, and Signal Processing*, pp. 3133-3136, May 1998.

[75] I. Hong and M. Potkonjak. "Behavioral Synthesis Techniques for Intellectual Property Protection," *36th ACM/IEEE Design Automation Conference Proceedings*, pp. 849-854, June 1999.

[76] P. Indyk, R. Motwani, P. Raghavan and S. Vempala, " Locality-Preserving Hash-ing in Multidimensional Spaces," *Proc. 29th ACM Symposium on the Theory of Computing*, 1997.

[77] D.S. Johnson, et al. "Optimization by simulated annealing: an experimental evaluation II. Graph coloring and number partitioning", *Operations Research*, Vol. 39, No. 3, pp. 378-406, 1991.

[78] S. Jung, R. Thewes, T. Scheiter, K. Goser, and W. Webber, "A Low-Power and High-Perfomance CMOS Fingerprint Sensing and Encoding Architecture," *IEEE Journal of Solid-State Circuits*, Vol. 34, No. 7, pp. 978-984, July 1999.

[79] D. Kahn. "The Codebreakers," The Macmillan Company, New York, NY, 1967.

[80] A.B. Kahng, J. Lach, W.H. Magione-Smith, S. Mantik, I.L. Markov, M. Potkonjak, P. Tucker, H. Wang and G. Wolfe. "Watermarking Techniques for Intellectual Property Protection," *35th ACM/IEEE Design Automation Conference Proceedings*, pp. 776-781, June, 1998.

[81] A.B.Kahng, S.Mantik, I.L.Markov, M.Potkonjak, P.Tucker, H.Wang and G.Wolfe. "Robust IP Watermarking Methodologies for Physical Design," *35th ACM/IEEE Design Automation Conference Proceedings*, pp. 782-787, June 1998.

[82] A.B. Kahng, D. Kirovski, S. Mantik, M. Potkonjak, and J.L. Wong. "Copy Detection for Intellectual Property Protection of VLSI Design", *IEEE/ACM International Conference on Computer Aided Design*, pp. 600-604, November 1999.

[83] R.M. Karp and M.O. Rabin, " Efficient randomized pattern-matching algo-rithms," *Technical Report*TR-31-81, Aiken Computation Laboratory, Harvard, 1981.

[84] M. Keating and P. Bricaud, "Reuse Methodology Manual for System-on-a-Chip Designs," Kluwer Academic Publishers, 1998.

[85] S. Khanna and F. Zane, "Watermarking Maps: Hiding Information in Structured Data," *(SODA'00)* pp. 596-605, 2000.

[86] B. W. Kernighan and S. Lin, "An Efficient Heuristic Procedure for Partitioning Graphs", *Bell System Tech. Journal* 49 (1970), pp. 291-307.

[87] K. Keutzer, " DAGON: Technology Binding and Local Optimization by DAG Matching," *Proc. ACM/IEEE Design Automation Conference,* 1987, pp. 341- 347.

[88] D.Kirovski and M.Potkonjak. "Efficient Coloring of a Large Spectrum of Graphs," *35th ACM/IEEE Design Automation Conference Proceedings*, pp. 427-432, June 1998.

[89] D. Kirovski, Y. Hwang, M. Potkonjak, and J. Cong. "Intellectual Property Protection by Watermarking Combinational Logic Synthesis Solutions," *IEEE/ACM International Conference on Computer Aided Design*, pp. 194-198, November 1998.

[90] D. Kirovski, D. Liu, J.L. Wong, and M. Potkonjak. "Forensic Engineering Techniques for VLSI CAD Tools", *37th ACM/IEEE Design Automation Conference Proceedings*, pp. 581-586, June 2000.

[91] J. Kleinberg, Y. Rabani, and É. Tardos, "Fairness in Routing and Load Balancing," *40th Annual Symposium on Foundation of Computer Science*, pp. 568-578, October 1999.

[92] D. E. Knuth, J. H. Morris and V. R. Pratt, " Fast Pattern Matching in Strings," *SIAM Journal on Computing* 6(2), 1977, pp. 323-350.

[93] E. Koch and J. Zhao, "Toward robust Hidden Image Copyright labeling," *Proceedings 1995 IEEE Workshop on nonlinear Signal and Image Processing*, pp. 452-455, June 1995.

[94] C.M. Koen Jr. and J.H. Im, "Software Piracy and Its Legal Implications," *Information and Management*, Vol. 31, pp. 265-272, 1997.

[95] R. A. Krutar, " Conversational Systems Programming (or Program Plagiarism Made Easy)," *Proc. 1st USA-Japan Computer Conference*, Oct. 1972, pp. 654- 661.

[96] A. Kündig, R.E. Bührer, and J. Dähler (Eds.) "Embedded Systems," Springer-Verlag, 1986.

[97] J. Lach, W.H. Mangione-Smith, and M. Potkonjak. "FPGA Fingerprinting Techniques for Protecting Intellectual Property," *Proceedings of the IEEE 1998 Custom Integrated Circuits Conference*, pp. 299-302, May 1998.

[98] J. Lach, W.H. Mangione-Smith, and M. Potkonjak. "Signature Hiding Techniques for FPGA Intellectual Property Protection," *IEEE/ACM International Conference on Computer Aided Design*, pp. 186-189, November 1998.

[99] J. Lach, W.H. Mangione-Smith, and M. Potkonjak. "Robust FPGA Intellectual Property through Multiple Small Watermarks," *36th ACM/IEEE Design Automation Conference Proceedings*, pp. 831-836, June 1999.

[100] T. Larrabee. "Test Pattern Generation Using Boolean Satisfiability," *IEEE Transactions on Computer AIded Design*, Vol. 11, No. 1, pp. 4-15, January 1992.

[101] C. Lee, M. Potkonjak, and W.H. Mangione-Smith. MediaBench: a tool for eval-uating and synthesizing multimedia and communications systems. International Symposium on Microarchitecture, pp.330-5, 1997.

[102] F. T. Leighton, "A Graph Coloring Algorithm for Large Scheduling", *Algorithms.Journal of Res.Natl.Bur.Standards*, Vol. 84, pp. 489-500, 1999.

[103] S.H. Low and N.F. Maxemchuk, "Performance Comparison of Two Text Marking Methods," *IEEE Journal on Selected Areas in Communications*, Vol.16, No.4, pp. 561-572, April 1998.

[104] S. Lu, V. Bharghavan, and R. Srikant, "Fair Scheduling in Wireless Packet Networks," *IEEE/ACM Transactions on Networking*, Vol. 7, No. 4 pp. 473-489, August 1999.

[105] I.R. Mackintosh, "Intellectual Property Protection White Paper: Schemes, Alternatives and Discussion Version 1.0", Virtual Socket Interface Alliance, January 2001.

[106] U. Manber, " Finding Similar Files in a Large File System," *Proc. Winter USENIX Conference*, 1994, pp. 1-10.

[107] J.P. Marques-Silva and K.A. Sakallah. "GRASP – A New Search Algorithm for Satisfiability," *IEEE/ACM International Conference on Computer Aided Design*, pp. 220-227, 1996.

[108] J.P. Marques-Silva and K.A. Sakallah. "Robust Search Algorithms for Test Pattern Generation," *Digest of Papers, 27th Annual International Symposium on Fault-Tolerant Computing*, pp. 152-161, June 1997.

[109] J.P. Marques-Silva and K.A. Sakallah, "Boolean Satisfiability in Electronic Design Automation", *37th ACM/IEEE Design Automation Conference*, pp. 675-680, June 2000.

[110] D.F.McGahn. "Copyright infringement of protected computer software: an analytical method to determine substantial similarity", *Rutgers Computer and Technology Law Journal,* Vol. 21, No. 1, pp. 88-142, 1995.

[111] P. McGeer, A. Saldanha, P.R. Stephan, R.K. Brayton, and A.L. Sagiovanni-Vincetelli. "Timing Analysis and Delay-Test Generation Using Path Recursive Functions," *IEEE/ACM International Conference on Computer Aided Design*, pp. 180-183, November 1991.

[112] A.J. Menezes, P.C. van Oorschot, and S.A. Vanstone, "Handbook of Applied Cryptography," CRC Press LLC, 1996.

[113] H. Morimura, S. Shigematsu, and K. Machida, "A Novel Sensor Cell Architecture and Sensing Circuit Scheme for Capacitive Fingerprint Sensors," *IEEE Journal of Solid-State Circuits*, Vol. 35, No. 5, pp. 724-731, May 2000.

[114] G.-J. Nam, K.A. Sakallah, and R. Rutenbar. "Satisfiability Based FPGA Routing," *Proceedings of the 12th International Conference on VLSI Design*, pp. 574-577, January 1999.

[115] R. Nelson, and R.J. Wilson (Editors) "Graph Colourings," Longman Scientific & Technical, Harlow,Essex, UK 1990.

[116] M. Niewczas, W. Maly and A. Strojwas, " A Pattern Matching Algorithm for Verification and Analysis of Very Large IC Layouts," *Proc. International Sym-posium on Physical Design*, 1998, pp. 129-134.

[117] M. M. Novak, *Correlations in Computer Programs, Fractals* 6(2), 1998, pp. 131-138.

[118] R. Ohbuchi, H. Masuda, and M. Aono, "Watermarking Three-Dimensional Polygonal Models Through Geometric and Topological Modifications," *IEEE Journal on Selected Areas in Communications*, Vol.16, No.4, pp. 551-560, April 1998.

[119] M. Ohlrich, C. Ebeling, E. Ginting and L. Sather, "SubGemini: Identifying Sub-Circuits Using a Fast Subgraph Isomorphism Algorithm," *Proc. ACM/IEEE De-sign Automation Conference*, 1993, pp. 31-37.

[120] A.L. Oliveira. "Robust Techniques for Watermarking Sequential Circuit Designs," *36th ACM/IEEE Design Automation Conference Proceedings*, pp. 837-842, June 1999.

[121] I. H. Osman and J. P. Kelly, eds., *Meta-Heuristics: Theory and Applications*, Kluwer, 1996.

[122] S. Pankanti and M.M. Yeung, "Verification Watermarks on Fingerprint Recongnition Retrieval," *Proceedings of the SPIE – The International Society for Optical Engineering*, Vol. 3657, pp. 66-78, January 1999.

[123] A. Parker and J. O. Hamblen, " Computer Algorithms for Plagiarism Detection," *IEEE Transactions on Education* 32(2), 1989, pp. 94-99.

[124] B. Pfitzmann. "Information Hiding Terminology", *The 1st International Information Hiding Workshop*, pp. 347-350, May 1996.

[125] B. Pfitzmann, and M. Schunter. "Asymmetric Fingerprinting," *Advances in Cryptology - EUROCRYPT'96, Proceedings of International Conference on the Theory and Application of Cryptographic Techniques. Maurer (Ed.)*, Springer-Verlag, pp. 84-95, 1996.

[126] B. Pfitzmann and M. Waidner, "Anonymous fingerprinting," *International Conference on the Theory and Application of Cryptographic Techniques Proceedings*, pp. 88-102, May 1997.

[127] C.P. Pfleeger, "Security in Computing (2nd Edition)," Prentice Hall PTR, February 2000.

[128] C. Podilchuk, W. Zeng. "Perceptual watermarking of still images," *IEEE Workshop on Multimedia Signal Processing*, pp. 363-368, 1997.

[129] P.W. Purdom Jr, and C.A. Brown. "Polynomial average-time satisfiability problems," *Inform. Sci.* 41, pp. 23-42, 1987.

[130] G. Qu. "Keyless Public Watermarking for Intellectual Property Authentication", *4th Information Hiding Workshop*, pp. 103-118, LNCS Vol. 2137, Springer-Verlag, April 2001.

[131] G. Qu. "Publicly Detectable Techniques for the Protection of Virtual Components", *38th ACM/IEEE Design Automation Conference Proceedings*, pp. 474-479, June 2001.

[132] G. Qu, and M. Potkonjak. "Analysis of Watermarking Techniques for Graph Coloring Problem," *IEEE/ACM International Conference on Computer Aided Design*, pp. 190-193, 1998.

[133] G. Qu, J.L. Wong, and M. Potkonjak. "Optimization-Intensive Watermarking Techniques for Decision Problems," *36th ACM/IEEE Design Automation Conference Proceedings*, pp. 33-36, June 1999.

[134] G. Qu, and M. Potkonjak. "Hiding Signatures in Graph Coloring Solutions," *3rd Information Hiding Workshop*, pp. 391-408, September 1999.

[135] G. Qu, J.L. Wong, and M. Potkonjak. "Fair Watermarking Techniques," *IEEE/ACM Asia and South Pacific Design Automation Conference*, pp. 55-60, January 2000.

[136] G. Qu, and M. Potkonjak. "Fingerprinting Intellectual Property Using Constraint-Addition," *37th ACM/IEEE Design Automation Conference Proceedings*, pp. 587-592, June 2000.

[137] M.K. Reiter and A.D. Rubin, "Anonymous Web Transactions with Crowds," *Communications of ACM*, Vol.42, No.2, pp. 32-38, February 1999.

[138] R.L. Rivest. "RFC 1321: the MD5 Message-Digest Algorithm," *Internet Activities Board*, April 1992.

[139] B. Rosenblatt, B. Trippe, and S. Mooney, "Digital Rights Management: Business and Technology," M&T Books, New York, NY 2002.

[140] N. Rudin, K. Inman, G. Stolvitzky, and I. Rigoutsos. "NA Based Identification," *BIOMETRICS personal Identification in Networked Society*, Kluwer, 1998.

[141] S. Rupley, "What's Holding up DVD?" *PC Magzine*, Vol. 15, No. 20, p. 34, November 1996.

[142] P. G. Salmon and R. J. Tracy. "Computer-Generated Computation Exercises", *Bahavior Research Methods and Instrumentation*, Vol. 7, No. 3, p. 307, 1975.

[143] R.G. van Schyndel, A.Z. Tirkel, and C.F. Osborne. "A digital watermark," *International Conference on Image Processing*, pp. 86-90, 1994.

[144] B. Selman, "Stochastic search and phase transitions: AI meets physics", *IJCAI*, pp. 998-1002, 1995.

[145] B. Selman, H.J. Levesque, and D. Mitchell, "A New Method for Solving Hard Satisfiability Problems", *National Conference on Artificial Intelligence*, pp. 440-446, 1992.

[146] B.Selman, and H.Kautz. "An Empirical Study of Greedy Local Search for Satisfiability Testing," *Proceedings of the 11th National Conference on Artificial Intelligence (AAAI-93)*, 1993.

[147] B. Selman, H. Kautz, and B. Cohen, "Local Search Strategies for Satisfiability Testing", *Cliques, Coloring, and Satisfiability: Second DIMACS Implementation Challenge*, 1993.

[148] B.Selman, H.Kautz, and D.McAllester. "Ten Challenges in Propositional Reasoning and Search," *Proceedings of the 15th International Joint Conference on Artificial Intelligence (IJCAI-97)*, pp. 50-54, 1997.

[149] N. Sherwani. *Algorithms for VLSI Physical Design AUtomation*, Kluwer Academic Publishers, 1995.

[150] N. Shivakumar and H. Garcia-Molina, " Building a Scalable and Accurate Copy Detection Mechanism ," *Proc. 1st ACM International Conference on Digital Li-braries*,1996, pp. 160-168.

[151] L.G. Silva, L. Silveira, and J. Marques-Silva, "Algorithms for Solving Boolean Satisfiability in Combinational Circuits," *Proceedings of the Design and Tests in Europe Conference*, pp. 526-530, March 1999.

[152] S. Singhe and F. J. Tweedie, " Neural Networks and Disputed Authorship: New Challenges," *Proc. International Conference on Artificial Neural Networks* London, 1995, pp. 24-28.

[153] P.R. Stephan, R.K. Brayton, and A.L. Sagiovanni-Vincetelli, "Combinational Test Pattern Generation Using Satisfiability," *IEEE Transactions on Computer Aided Design*, Vol. 15, No. 9, pp. 1167-1176, September 1996.

[154] J.P. Stern, G. Hachez, F. Koeune, and J.J. Quiquater, "Robust Object Watermarking: Application to Code", *3rd Information Hiding Workshop*, Lecture Notes in Computer Science, Vol. 1768, pp. 368-378, Springer-Verlag, 1999.

[155] R. H. Storer, S. D. Wu and R. Vaccari, "New Search Spaces for Sequencing Problems With Application to Job Shop Scheduling", *Management Science* 38 (1992), pp. 1495-1509.

[156] P.H. Sullivan, S.P. Harrison, G.N. Keeler, and J. Villella, "The Value and Management of Intellectual Assets Version 1.0", Virtual Socket Interface Alliance, March 2002.

[157] M.D.Swanson, B.Zhu, B.Chau, and A.H.Tewfik. "Object-based transparent video watermarking," *IEEE Workshop in Multimedia Signal Processing*, pp. 369-374, 1997.

[158] M.D. Swanson, B. Zhu, and A.H. Tewfik, "Robust Data Hiding for Images," *Proceeding IEEE Digital Signal Processing Workshop*, pp. 37-40, September 1996.

[159] M.D. Swanson, B. Zhu, and A.H. Tewfik, "Multiresolution Scene-Based Video Watermarking Using Perceptual Models," *IEEE Journal on Selected Areas in Communications*, Vol.16, No.4, pp. 540-550, April 1998.

[160] R. Thisted and B. Efron, "Did Shakespeare Write a newly discovered Poem?" *Biometrika*, Vol. 74, pp. 445-455, 1987.

[161] K. L. Verco and M. J. Wise, " Plagiarism a la Mode: a Comparison of Automated Systems for Detecting Suspected Plagiarism," *Computer Journal* 39(9), 1996,pp. 741-750.

[162] N.R. Wagner. "Fingerprinting," *Proceedings of the 1983 Symposium on Security and Privacy, IEEE Computer Society*, pp. 18-22, 1983.

[163] D.de Werra. "An Introduction to Timetabling", *European Journal of Operations Research*, Vol. 19, pp. 151-162, 1985.

[164] G. Wolfe, J.L. Wong, and M. Potkonjak. "Watermarking Graph Partitioning Solutions", *38th ACM/IEEE Design Automation Conference Proceedings*, pp. 486-489, June 2001.

[165] R.B. Wolfgang, C.I. Podilchuk, and E.J. Delp, "Perceptual Watermarks for Digital Images and Video," *Proceedings of the IEEE*, Vol.87, No.7, pp. 1079-1107, July 1999.

[166] M.M.Yeung, F.C.Mintzer, G.W.Braudaway, and A.R.Rao. "Digital watermarking for high-quality imaging," *IEEE Workshop on Multimedia Signal Processing*, pp. 357-362, 1997.

[167] H. Zhang, "Service Discipines for Guaranteed Performance Service in Packet-Switching networks," *Proceedings of the IEEE*, Vol. 83, No. 10, pp. 1374-1396, October 1995.

[168] H. Zhang "SATo: AN Efficient Propositional Prover," *Proceedings of International Conference on Automated Deduction*, July 1997.

[169] International Technology Roadmap for Semiconductors, *http://public.itrs.net*, 2001.

[170] Virtual Socket Interface Alliance. "System Chip Letter," Issue 2, Summer 1998.

[171] " The GNU awk program ", Available by anonymous FTP from prep.ai.mit.edu.

[172] http://www.cs.ualberta.ca/~joe

[173] http://aida.intellektik.informatik.th-darmstadt.de/ hoos/SATLIB/

[174] http://dimacs.rutgers.edu/

[175] http://mat.gsia.cmu.edu/COLOR/instances.html

FSC
www.fsc.org
MIX
Papier aus verantwortungsvollen Quellen
Paper from responsible sources
FSC® C105338

If you have any concerns about our products,
you can contact us on
ProductSafety@springernature.com

In case Publisher is established outside the EU,
the EU authorized representative is:
Springer Nature Customer Service Center GmbH
Europaplatz 3, 69115 Heidelberg, Germany

Printed by Libri Plureos GmbH
in Hamburg, Germany